Researching Practitioner Inquiry as Professional Development

Rose M. Pringle

Researching Practitioner Inquiry as Professional Development

Voices from the Field of Science Teaching

Springer

Rose M. Pringle
College of Education, School of Teaching and Learning
University of Florida
Gainesville, FL, USA

ISBN 978-3-030-59549-4 ISBN 978-3-030-59550-0 (eBook)
https://doi.org/10.1007/978-3-030-59550-0

This Springer imprint is published by the registered company Springer Nature Switzerland AG
The registered company address is: Gewerbestrasse 11, 6330 Cham, Switzerland

This book is dedicated to George, George Junior, and Johanna as well as Renee and Mark.

Keegan and Kaila, may you always have the best teachers!

Preface

There are numerous texts on the market that describe the process and procedures for conducting practitioner or teacher inquiry. There are also many texts on professional development (PD) in general that have, as the primary goal, the promotion of teachers' learning and improved pedagogical practices. This book, *Researching Practitioner Inquiry as Professional Development: Voices from the Field of Science Teaching,* strategically presents the authentic voices of middle school science teachers validating the use of teacher inquiry in facilitating their learning and in shaping their classroom practices. Practitioner or teacher inquiry is teachers' intentional involvement as researchers steeped in the systematic study of their own teaching practices. The impetus for making inquiries into their own practices emerges from a place of concern for students' learning and the teachers' desire to make the necessary pedagogical adjustments.

This book is about middle school science teachers, their engagement in a multifaceted PD program, and the deliberate practice of practitioner inquiry as a tool to support their learning. In the broader context, the book lays bare the role of science education and its development over the years to contribute to the development of a scientifically literate society. A society in which its citizens have a good understanding of the nature of science and grasp the local, national, and global implications of scientific issues such as climate change and renewable energy. Science education is also expected to ensure the availability of a continued pipeline to careers in science and engineering. At this current juncture, and in the face of new societal expectations, science education has had to make the necessary adjustments. Changes in society and the new wave of reform being spurred on, however, are not occurring in a vacuum. Hence, the book introduces and briefly discusses major reforms in science education. The discussion begins with the era of Sputnik and ends with the current reform as described in *A Framework for K-12 Science Education: Practices, Crosscutting Concepts, and Core Ideas.* What follows the historical discussion is a description of the PD program, including the NSF-sponsored *Teacher Institute for the 21st Century*, and other complementary workshop activities. The program described in Chap. 3 supported the preparation of middle school science teachers who would, as agents of change, lead district-wide transformation of science teaching. Four years later, the program was scaled and now being sustained as a one-year professional learning certificate program for third to eighth grade teachers of science.

Chapter 4 describes the intense process of introducing the group of teachers to practitioner or teacher inquiry and to its acceptance as being viable within the context of education research. The chapter is described as intense because the course and the process of conducting the teacher inquiry challenged the teachers' beliefs about research, their engagement as researchers, and the extent to which their work would have credence within their schools and districts. The efforts of the PD were to ultimately increase students' science achievement by positioning the group of teachers as agents of change in the transformation of science teaching. The program therefore embraced the current reform efforts in science education and was guided by a conceptual framework that situated teachers as learners in their classrooms as they deepen both disciplinary content knowledge and science-specific teaching practices. The teachers' introduction to practitioner inquiry as research, their struggles in accepting the process, and how they eventually conceptualized their engagement in inquiry as PD are carefully documented. Chapter 4 is important to understanding the process of developing a culture of practitioner inquirers among practicing teachers and also sets the context leading to the voices of the teachers presented in the following three chapters.

Chapters 5, 6 and 7 give credence to the teachers' experiences. The teachers, immersed in the cycle of inquiry, illuminated the complex and challenging nature inherent in teaching. Their stories provide a first-hand account of the potentials and possibilities of practitioner inquiry as PD and teachers as knowledge producers. The narratives are presented around three major themes that emerged from the analysis of the issues of practice addressed by the teachers. Within each theme, three representative narratives are presented. Each chapter ends with a critical discussion of the importance of the knowledge garnered from the teachers' experiences and the link between theory and practice.

In Chap. 8, the final chapter, I position practitioner inquiry as an element of PD and discuss the impact of the process on teachers' learning as indicated in their reflections and response to a formal interview. The book is about elevating teachers' voices and giving credence to their positions as knowledge generators. Hence, in continuing this theme, their reflections and their perspectives are used to frame the value-added components of practitioner inquiry within the context of their learning. During the process of conducting the inquiry, teachers systematically and intentionally observed and reflected on their teaching with specific intent to make adjustment toward improved practice. I present the teachers' reflections on their learning, the ways their practices were impacted by the newly constructed knowledge, and how, in some instances, they were forced to confront policies and systemic issues. Teachers were aware of the issues in the broader systems that inhibited change, but over time, they collectively began to conceptualize practitioner inquiry as a tool that can be incorporated and harnessed to support effective practices in their local contexts.

The teachers agreed that the standard-driven science curriculum that included learning progression, integration of science content and skills, and well-defined instructional practices for supporting students' learning provided ease during its enactment. In planning for and conducting the practitioner inquiries, teachers therefore focused on vexing issues in their own contexts that were negatively impacting students' learning. In the process, teachers built closer and stronger relationship with their students, developed a range of teaching strategies specifically related to issues of learning, and became empowered as they adjusted their instructions. The book ends with making the case for elevating practitioner inquiry to the status of being an element of effective PD. When teachers as learners are driven by the need to increase students' achievements, they confront their realities and make the necessary pedagogical adjustments in their local contexts. This is the essence of teacher learning and the beginning of changing the culture of learning in schools – a focus on improving students' achievement from the inside.

Acknowledgments

The University of Florida Unites Teachers to Reform Education in Science (U-FUTuRES) project was funded by the National Science Education Math/Science partnership program, Award # 1050166. The book, *Researching Practitioner Inquiry as Professional Development: Voices from the Field of Science Teaching,* focuses on one aspect of the 5-year federally funded–program – teachers conducting practitioner inquiry. The program was a comprehensive endeavor. I acknowledge that the successes we have achieved are attributable to the dynamic team of individuals who contributed to the planning and enactment of the program throughout the years.

First, I acknowledge Dr. Lynda Hayes, who, as the principal investigator and in her positions as a school and district administrator and an affiliate university professor, occupied a unique position from the program's conception through to its sustainability. Her wisdom and knowledge of the nature of schools, administration, and partnerships were superb in the creation, alignment, and achievement of the vision of U-FUTuRES.

Dr. Jenn Mesa was the program's associate for science education. Her passion for science teaching and learning and her tireless collaborative effort supported the translation of the project plans into meaningful learning activities for the teachers. Jenn, your involvement in the program will always be greatly appreciated.

I am grateful for the managerial and organization skills and attention to details that Dr. Leela Kumaran brought to the program in her role as program associate. Her knowledge and expertise were priceless!

I extend gratitude to Dr. Natalie King, Dr. Joanne LaFramenta, Dr. Tim Barko, and Mrs. Erin Mistry for the extensive travels to observe classrooms, offer support in course development, and the valuable contributions to the different phases of the program.

Dr. Karen Kilgore, thanks for your assistance in the final research efforts of the program. Your skills in data collection and your assistance in analysis contributed greatly to the articulation of the teachers' voices.

The program was for middle school science teachers. I thank all 37 teachers who participated in the initial program. I also extend many thanks to the nine teachers whose voices have been amplified in the telling of their stories: Mayra, Allison,

Anthony, Brittni, Jennifer, Janet, Ruth, Sean, and Sandra. May you all continue to be stellar in the field of science teaching.

This book is based upon work supported by the National Science Foundation under Grant No. 1050166. The opinions, findings, and conclusions or recommendations expressed in this material are those of the author and do not necessarily reflect the views of the National Science Foundation.

About the Book

With every new wave of reforms in science education, there is a need for effective professional development (PD). This book, *Researching Practitioner Inquiry as Professional Development: Voices from the Field of Science Teaching,* has at its core the experiences of middle school science teachers engaged in a professional development (PD) program grounded in current science reform. The book describes the PD program that had as its goal the transformation of science teaching practices and increase in students' learning. The teachers were immersed in their classrooms as they developed and practiced the new knowledge and skills. The practitioner inquiry served to satisfy the university's requirement for the graduate program but also afforded the teachers another strategy to continue their learning and to gain more insights into their teaching. As reflective practitioners, the teachers observed, reflected on, and adjusted their practices in support of science learning. The book shares nine teachers' authentic experiences, thus allowing us a window into their classrooms. The value of practitioner inquiry as a genre of educational research is discussed and the teachers' voices provide a framework from which other educators can begin to reflect on their own problems of practice in working with inservice teachers. Researching the teachers' inquiry brings to the fore the need for practitioner inquiry to occupy a place among the elements of effective PD.

Contents

About the Author

Rose M. Pringle is an associate professor of science education in the School of Teaching and Learning at the University of Florida. As a science educator, her research focuses on interrelated themes along the continuum of science teacher education: pre-service teachers' positionality as science learners, science-specific pedagogies of both prospective and practicing science teachers, and the translation of these practices into equitable inquiry-based science experiences for all learners. Currently, she is exploring elements of effective science instruction consistent with current reform in science education and pedagogical content knowledge as frameworks for gauging the practices of science teachers. Of particular interest is her quest to increase the participation of underrepresented minorities, especially girls of African descent, in science and science-related careers. Her work has revealed that school-wide policies and teachers' autonomous decisions impact the regularity and nature of science instruction, and teachers do not always conceptualize African American girls as science achievers positioning them in negative ways. Therefore, Rose is examining how teachers develop knowledge about teaching and learning science and how engaging in practitioner or teacher inquiry will increase their cultural competence and strengthen their science-specific teaching practices. She posits that teachers' attention to culture and the examination and adjustment of their practices to accommodate diverse learners will increase students' achievement in science.

Abbreviations

3D	Three Dimensional
AAAS	American Association for the Advancement of Science
ASD	Autism Spectrum Disorders
CER	Claims, Evidence, and Reasoning
CK	Content Knowledge
CL	Cooperative Learning
DLDS	District Leadership Development Series
ESE	Exceptional Student Education
IEP	Individual Education Plan
IHE	Institution of Higher Education
IQWST	Investigating and Questioning Our World Through Science and Technology
K-12	Kindergarten to Twelfth Grade
KWL	Know-Want-Learn
LD	Learning Disabilities
LEA	Local Education Agencies
MAE	Master of Arts in Education
MSP	Math/Science Partnership
NASEM	National Academies of Sciences, Engineering, and Medicine
NEFEC	Northeast Florida Educational Consortium
NGSS	Next Generation State Standards
NRC	National Research Council
NSF	National Science Foundation
NSTA	National Science Teacher Association
OHI	Other Health Impairments
PCK	Pedagogical Content Knowledge
PCK	Pedagogical Content Knowledge
PD	Professional Development
PKY	P.K. Yonge Developmental Research School

STEM	Science, Technology, Engineering, and Mathematics
STLI	Science Teacher Leadership Institute
U-FUTuRES	University of Florida Unites Teachers to Reform Education in Science
U.S.	United States

List of Figures

List of Tables

Part I
Setting the Stage

Chapter 1
Introduction

With every new wave of reforms, there is a need for professional development (PD). The publication of A Framework for K-12 Science Education in (2012) was deemed an important first step in reforming science education for the twenty-first century. The efforts described in the reform document were the result of years of research, intent on revitalizing science education, and increasing its relevance. The history of science education in the United States (U.S.) has taught us that elements of reform and the duration of their effectiveness are contingent on contemporary socio-political and economic outlook and the advancements of knowledge. In the past, major changes in science education occurred in response to historical and social circumstances (Bybee, 1993). The current framework acknowledges the foundation laid by earlier reform efforts and the strong research that ushered in the development of standards in the mid twentieth century. In this era of standard-driven educational system, standards are embraced as the set of educational aspirations positioned to guide curricular and instructional activities. As descriptions of what students are expected to know and be able to do, standards have been afforded much prominence in education. The national science education standards, though not adapted by all states, provide a set of goals and guidelines for science learning in the U.S.

The documented shortcoming in achieving society's goals of the development of a scientifically literate population and a workforce with twenty-first century skills created the need for the current wave of transformation in science education. The appeal for a better approach to science education was also borne out of the revolutions in science and science knowledge and in other related disciplines such as technology, engineering, and mathematics. The reform effort generated to improve K-12 science teaching and learning has at its core the organization of selected conceptual knowledge frameworks coordinated around core ideas, crosscutting concepts, and learning progression. These frameworks have implications for core knowledge learning, instruction, and assessment.

While the reform ideas are generated by educators and other stakeholders, classroom teachers are central to their translation into practice. Inservice teachers in the current school system are seasoned in their approach to teaching science. They are

R. M. Pringle, *Researching Practitioner Inquiry as Professional Development*,
https://doi.org/10.1007/978-3-030-59550-0_1

guided by their beliefs as to what constitutes best teaching strategies and the organization and nature of their curriculum. To effect changes as envisioned in the reform, teachers need to be sufficiently prepared with the depth of understanding of the content knowledge and the science-specific teaching strategies to enact the new approach. This certainly heightens the necessity for PD programs to prepare teachers in ways that will ensure that the envisioned conceptions of reform efforts are ingrained in their belief systems to guide and impact their practices. This was the vision of the *University of Florida Unites Teachers to Reform Education in Science (U-FUTuRES)* project – a comprehensive PD program. With a focus on students' learning, the activities increased the content knowledge of middle school teachers, engaged them in inquiry-based science as learners and allowed them to practice, reflect, and refine their practices within the supportive learning community developed and sustained within the program.

The program was conceptualized and initiated within a partnership that included schools and districts populated by poor and rural students including African Americans and Hispanics and for whom participating in science careers was not a foreseeable reality. Many of the schools were not meeting the expectations of the science standards and their history of low science achievement was not only an issue for the future of the students but was a factor in the accountability system established by the state. While the motivation for teachers' learning is usually from within (Simon & Campbell, 2012), in this program, they were selected for participation after being recommended by their administrators. However, from as early as their initial interviews, before being matriculated into the university system, they all expressed that they accepted the program's long-term vision to impact the academic trajectory of their populations of learners.

The PD program included a job-embedded graduate degree program and complementary workshop activities aligned with the formal courses. A core component was the preparation of the teachers to enact the middle school science curriculum grounded in the elements of the current reform in science education. That is, the program strategically and intentionally facilitated the teachers developing a strong understanding of the scientific ideas and practices as outlined in the reform and through the enactment of the curriculum translated their learning into practices. Informed by social constructivism, the program facilitated teachers' learning through active engagement and as reflective practitioners reconstructed their existing understanding of science teaching and learning. That is, teachers moved beyond dissonance toward transformed teaching practices aligned with the current reform in science education.

The immediate context for this book is a Math Science partnership funded by the National Science Foundation (NSF). The funding period of 5-years supported two cohorts of 37 middle school teachers in a *Teacher Institute for the twenty-first Century* that included a series of comprehensive PD activities and the completion of a 2-year job-embedded master's degree in science education. The book, Researching *Practitioner Inquiry as Professional Development: Voices from the Field of Science Teaching,* has at its core, the experiences of science teachers in the PD program as they share the authentic learning experiences that emerged during the practitiioner inquiry. The teachers, immersed in their sites of practice, and as deliberative intellectuals, engaged in systematic and intentional study of their teaching toward improving students learning (Dana & Yendol-Hoppey, 2019). The nine stories presented in chapters five through seven include the teachers' informed perspectives as they observed, reflected on their actions, and then adjusted their teaching to enhance students' science learning. The book does not present a sanitized version of the teachers' stories. It describes the full PD program in which the teachers were prepared to be agents of transformation within the current reform in science education. Through the sharing of the teachers' narratives, we are provided a window into real-time occurrences in science teaching and learning in middle school classrooms.

The practitioner inquiry was one component of the PD program. The authentic teachers' voices represent the culmination of their introduction to, and completion of their first practitioner inquiry. The book begins with a historical discussion providing insights into past experiences that have shaped the current science education culture. In the midst of current socio-political and cultural shifts, an understanding of the history of science education is important to appreciate the significance of the interaction between science and society and the emergence of priorities over the years. What follows the historical discussion, is a description of the PD program and the course work that introduced and scaffolded the teachers into and through the cycle of teacher inquiry. The content of the book is not about the author's experiences as a science educator whose research is in teacher education. It is about teachers' learning that emerged from inquiring into their issues of practice, their posture as reflective practitioners, and the generation of valuable knowledge about teaching and learning science. Thus, three major areas of their concerns are presented in chapters five through seven. While I have selected literacy in science learning, cultural competence, and development of metacognitive skills to be presented here, other areas of concerns explored in the teacher inquiries were, gender and science learning, gifted but low achieving in science, and migrating populations of learners. The book ends with a critical discussion of practitioner inquiry as a force in supporting teacher learning as garnered from the teachers' reflections. Finally, the case is made that all PD activities should include practitioner inquiry. However, as an educational movement and as a research genre (Cochran-Smith & Lytle, 2009), teachers should be introduced and scaffolded through a formal job-embedded learning experience that includes the cyclic process from observation through to transformed practices and leading to further observation.

1.1 A Resource in Inservice Teacher Education

The book can be interpreted as a historical documentation of one component of a twenty-first century PD; an attempt to reveal, through the voices of science teachers, issues of practice in real classrooms; or a framework from which other educators can begin to reflect on their own problems of practice in working with inservice teachers regardless of the disciplines. The goals for those who read and use this book are to (1) develop an understanding of science education, its relationship to science, and the impact of social, global, and economic occurrences on its development; (2) provide a window into middle school classrooms through the authentic voices of science teachers engaged in practitioner inquiry; and (3) bring to the fore the power of practitioner inquiry as a tool in PD to support the development of reflective practitioners. The book is intended for a wide audience including science educators, school and district administrators, and policymakers. The information is also applicable to educators whose discipline may not be science education but who are engaged with teachers in efforts to increase students' achievement. The following are recommendations for use of the book by educators in higher education and educators and administrators whose interests are in enhancing the professional learning of all teachers.

1.1.1 For Science Teacher Educators: The Making of Scholars

Foregrounding the aspects of the history of reform in science education provides insight into a field that philosophically has been engaged in a process of legitimizing its place in an evolving world. While there are numerous texts on the history of science education, the information presented within the context of a PD program is pertinent for graduate students being prepared for academia. The historical discussion lays the foundation for understanding how past experiences have shaped the current science education culture and its relationship to the society's expectation. In the midst of current socio-political and cultural shifts, an understanding of the history of science education not only highlights the significance of the interactions between science and society and the emergence of priorities over the years but will make for scholarly discussions in graduate seminars. In addition, understanding the historical antecedents to the current reform efforts and the implication for PD has the potential to generate research agendas to further promote the achievement of the long-term goals of the discipline.

1.1.2 For Inservice Teacher Educators and School Administrators

Regardless of the discipline of focus, the information provided in the text can be used to shape conversations around the context and opportunities for more powerful professional teacher learning. The experiences of these teachers reveal the extension of their role as they developed the professional dispositions of life-long learning, reflective and mindful teaching, and self-transformation (Mills, 2000). As scholars, they developed knowledge about their teaching and students learning – a testament to the scope at which teachers, as reflective practitioners can learn from their experiences and make the necessary changes to their practices. To achieve such, however, the information contained in the book can guide the development of an approach that will reframe the nature of PD from being static in time to one that empowers teachers as continuous learners. It is this empowerment, emerging from the practitioner inquiry that will facilitate continued learning beyond any given duration.

As administrators seek to encourage environments in which teachers contribute to school change, the book provides insight into a component of PD that allows teachers to identify the issues associated with their practices and simultaneously study their teaching, their students and themselves. With the focus on improved practices and student learning, the powerful voices of the teachers and the impact of teacher inquiry on their learning are areas that can foster teachers' involvement in school change (Elmore, 2004). Furthermore, this text becomes a unique resource for initiating and or supporting learning communities (DuFour, 2004) on school campuses. The history, description of the program, and the narratives are precursors to teachers making connections to the lived experiences presented by their peers. The final chapter presents a succinct discussion of the knowledge garnered from researching practitioner inquiry and its potential to be a catalyst in school change.

References

Bybee, R. W. (1993). *Reforming science education. Social perspectives & personal reflections*. New York, NY: Teachers College Press.

Cochran-Smith, M., & Lytle, S. L. (2009). *Inquiry as stance: Practitioner research for the next generation*. New York, NY: Teachers College Press.

Dana, N. F., & Yendol-Hoppey, D. (2019). *The reflective educator's guide to classroom research: Learning to teach and teaching to learn through practitioner inquiry* (4th ed.). Thousand Oaks, CA: Corwin.

DuFour, R. (2004). "What is a" professional learning community? *Educational Leadership, 61*(8), 6–11.

Elmore, R. F. (2004). *School reform from the inside out: Policy, practice, and performance*. Cambridge, MA: Harvard Education Press.

Mills, G. (2000). *Action research: A guide for the teacher researcher*. Upper Saddle River, NJ: Merrill/Prentice-Hall.

Simon, S., & Campbell, S. (2012). Teacher learning and professional development in science education. In B. Fraser, K. Tobin, & J. M. R. Campbell (Eds.), *Second international handbook of science education* (pp. 307–321). Dordrecht, Netherlands: Springer.

Chapter 2
Reforms in Science Education: A Response to Changing Societal Contexts

Reforms in science teaching and learning have often emerged in response to major socio-political and economic occurrences. The launching of Sputnik by the Soviet Union in the 1950s was viewed as a challenge to the scientific and technological prowess of the United States (U.S.) (Stine, 2008). This was a new age in geopolitics. An age in which science education was expected to play an important role in response to global challenges in science, technology, and the advancement of nation states. The launch prompted a rapid national response that resulted in the mobilization of new federal policies, educational programs, and a call for the renewing of intellectual rigor to school science programs (DeBoer, 1991; Rudolph, 2002). The need for a vibrant, science and mathematics enriched learning environment, raised the concerns of scientists and spurred their input into education. The intellectual contest with the Soviets provided the opportunity for scientists to become involved in revitalizing the science curriculum (Rudolph, 2002). A tide of new approaches to science curricular development influenced by these scientists, lead to initiatives promoting science teaching and learning that emphasized the logical structure of the disciplines and the processes of science (DeBoer, 1991; Rudolph, 2002).

Science education in the immediate post-Sputnik era, saw an emphasis on the learning of process skills such as observing, inferring, and experimenting. Later, in the 1990s, the National Research Council (NRC, 1996a), released the National Science Education Standards (NSES), which, even today, are embraced as an important document in giving directions to setting the ideals for teaching and learning science. The central goal of the standards was the establishment of the major purpose of science education which was the development of scientific literacy for all students, not just those destined for careers in science and engineering. Embedded in this goal was the expectation that all students would be given equitable science learning opportunities during their K-12 schooling. Thus, the emergence of the phrase, *Science for all students*. The principles of the standards included an emphasis on teaching science in ways that reflected how science itself is conducted by scientists – emphasis on inquiry as a way of achieving knowledge and

R. M. Pringle, *Researching Practitioner Inquiry as Professional Development*,
https://doi.org/10.1007/978-3-030-59550-0_2

understanding about the world (National Research Council (NRC), 1996). Embedded within the reform effort, was the distinction between content standards that identified the science ideas to be learned and inquiry standards that established what students should be able to do (NRC, 2000). In this dispensation of science teaching and learning, the goals of inquiry became clarified by attempts to teach science as practice (Abd-El-Khalick, Bell, & Lederman, 1998; Lehrer & Schauble, 2006). It was during this time in the development of science education that attention given to the process skills in science learning became heightened.

The national standards directed a vision that embraced science as practice within the context of an inquiry approach to science teaching and learning. This notion of inquiry, however, was not defined operationally in the accompanying reform documents. The lack of a consensus lead to varying conceptualizations and translations in the science classrooms that did not accurately represent the vision of inquiry as envisioned in the reform. As the standards' movement became central in the reformed educational systems, a plethora of K-12 science programs emerged that had implications for teaching, learning and assessment, and the preparation of teachers. In addition, on a political level, the standards-based teaching ushered in the era of tests and accountability, which overtime, have become a constraint on the achievements for which it was instituted to support.

By the dawn of the twentieth century, the influence of the rapid changes in approaches and methods brought about in response to the Sputnik era was far removed from the evolving discipline, science education. So too, were the issues and concerns in education that initiated the report, "A Nation at Risk" (National Commission on Excellence in Education, 1983). Notably, at this juncture, the lingering challenges from the past centuries remained and were very evident in the current state of science education. Seventeen years after the emergence of the "Science for all students," the science education community was again faced with the failure of meeting society's expectations. Science education was still plagued by the low level of scientific literacy, less than adequately prepared citizens to respond to the prevalence of science in the society, and the need to prepare a scientifically trained workforce (NRC, 2007). In addition, critics cited the inadequacies of the education system to provide learning experiences that engaged all students in the enterprise of science.

A review of the past discussions of the early 1900s reveals some interesting conversations and lamentations about the state of K-12 science education. Specifically, its role in securing the development of knowledge, skills, and habits of mind leading to personally fulfilling and responsible lives. However, more than the individual self-fulfillment and the economic and natural interests were at stake. At stake, according to Rutherford and Ahlgren (1990), was attention to global challenges including but not limited to acid rain, shrinking of tropical forests and other sources of diverse species, and pollution of the environment. These conditions and conversations are eerily similar to occurrences in our current system and so too are the efforts for remediation. In Bybee's (1993) reflections on science education, he posited that increasing the development of scientific literacy would require fundamental changes in the approach to science curricula. Some of his suggestions included a reduction in the volume of content knowledge with a focus on a few major concepts and skills; greater emphasis on interdisciplinary learning with a softening of the rigid boundaries of the individual science disciplines; and the enactment of programs that were conceptually sound and contained procedural integrity. Furthermore, in 1983, Bybee projected that in the next decade, the issue of equity would need attention to enhance the opportunities for historically underrepresented groups. Despite the standards call for a commitment to the provision of equitable learning experiences for all learners (NRC, 2007), under-representation in science by ethnic minorities remains a function of the nature of K-12 science education.

The practices of science teaching envisioned in the science standards for all and the conceptions of science as inquiry, became synonymous with science learning as the assimilation of discrete facts. In such practices, teaching was about covering science content that was "a mile wide and an inch deep," which, according to Schmidt, Burroughs, and Cogan (2013), supported a surface learning of the science content knowledge. Teaching with an emphasis on accumulation of facts did not allow students to experience science as a dynamic enterprise and in ways consistent with how scientists work in studying the world. With the science knowledge permeating every aspect of the modern world, science education was once again not meeting the need of the U.S. to secure economic dominance and maintain its competitiveness in the international arena. Furthermore, the impact of deficiencies in K-12 learning was being felt at both the personal and communal levels. There was now an obvious failure in the extent to which citizens on a whole were participating in public issues and prepared to make informed decisions. An ineffective science education program therefore became a signal for a new framework, accompanying standards, and a more intentional and inclusive approach to K-12 science teaching and learning.

2.1 Science Education and the Development of Twenty-First Century Skills

We live in a world that is profoundly being shaped by the advancing and ever-increasing volume of science knowledge. "This scientific advancement has fundamentally changed our beliefs about what it means to do science, to engage in scientific inquiry and to describe science as a way of knowing," (Duschl & Grandy, 2008, p. ix). These changes have had significant implications for the nature of science teaching and the demand for quality education in K-12 classrooms. The commitment to a program, that over time, embodies these new beliefs was also supported by a new understanding of learning informed by cognitive science research (Bransford, Brown, & Cocking, 1999). Amidst these changes, the goals of science education have remained consistent. The goals include the development of the intellectual skills of thinking and reasoning, practical goals of empowerment that will allow for participation in public policies and discourse, and the futuristic goals of creativity and innovation (DeBoer, 1991; Duschl & Grandy, 2008). The participation in public discourse has taken on greater meanings in the twenty-first century as issues such as preserving the environment, global warming, clean air, and generating energy efficiency are being hotly debated in the political realm. Furthermore, within the current economic, environmental, and social challenges, a good science education is important to secure gainful employment. Children with inadequate learning experiences are more likely as adults to lack the capabilities to be critical consumers of science information related to their everyday lives (U.S. DOE, 2000).

At a time when there is the quest to maintain economic and political dominance within the global marketplace, the shortage of knowledge, skills, and dispositions related to science education, is a cause for concern. The importance of science is widely recognized in both developing and developed nations in the context of global economy, advancing science and technology, and the global redistribution of skilled workforces. In the U.S., the federal report *Rising Above the Gathering Storm, Revisited: Rapidly Approaching Category 5.* (National Research Council, 2010), recommended some top actions to enhance the enterprise of science and technology toward securing America's competitiveness, prosperity, and security in the global community of the twenty-first century. The revitalization of science and mathematics within K-12 education was afforded the highest priority area. According to the report, an effective K-12 science and mathematics program is crucial in addressing the unease about long term trends and attending to the urgency in strategic and economic security.

The pursuit of scientific understanding, the application of engineering solutions, and technological innovations are all factors contributing to the economic future of nation states (NAS, 2007; Timms, Moyle, Weldon, & Mitchell, 2018). In the United Kingdom (UK), recent reports of the National Commission on Education and the Royal Society have both called for a revitalization of K-12 science and technology education in maintaining the nation's status as a leading economy. The report from the Royal Society acknowledges the important contribution of science and mathematics education to the intellectual development of its young people and the fostering of new cadres of professional in science, engineering and mathematics. Also, in Australia, a statement from the Science Priorities for Australian Innovation (2015) presents science as the "engine room" to drive the country's innovation and prosperity into the future. However, to deliver continued prosperity and solve grand challenges, the statement calls for a vibrant and well-planned and well-resourced science and mathematics education for all Australian students. Science with its specialized knowledge, skills, and dispositions, is clearly linked to the broad issue of human capital deficit and the generative and economic power across the globe.

2.1.1 A New Framework for Science Education

As in the past, the conversation around the crisis in science education in the twenty-first century is in part attributed to issues that seem to be persistent across time. One key issue is the science achievement gaps among U.S. K-12 populations when comparisons are made with their international peers. Another issue is the decline in the number of students choosing to study science or pursuing science-related careers and the failure of the education system to ensure the development of a scientifically literate society. The document, *A Framework for K-12 Science Education* was commissioned in direct response to the decline in science education and the lack of fundamental knowledge among the workforce especially in the areas of Science, Technology, Engineering, and Mathematics (STEM). The development of the framework was guided by research on teaching and learning (Donovan & Bransford, 2005), and an agreement reached among educators and scientists as to what constitutes foundational knowledge and skills for science learning. The framework articulated a set of expectations that informed the development of the new science standards and subsequently guided new approaches to science teaching, learning, and assessment. An approach, that according to Harris et al. (2015) builds upon the first generation of U.S. science standards but with some differences. Several features of the current framework set it apart from earlier reforms. Some of the new features include the identification of core ideas and crosscutting concepts, the systematic progression of content knowledge and science and engineering practices over the K-12 years of schooling, and attention to how learning occurs.

With a focus on students' proficiency and appreciation of science, the new science framework recommends that K-12 science be developed around three major dimensions (3D); disciplinary core ideas, crosscutting concepts, and science and engineering practices. In embracing "less is more," the 3D approach to science teaching and learning as proposed, seeks to foster the development of complex understanding. Students will be provided with a foundation on which abstract ideas can be linked with prior understanding, allowing them to connect ideas and deepen their conceptual understanding in the face of new information. Guided by our current understanding of how learning occurs, students as active learners are allowed to reflect on and challenge their prior understanding in the presence of new experiences. This approach removes earlier practices and enactment of science curricular that prioritized the learning of snippets of science information over the process of sense making. It now lays the foundation for the appreciation that science knowledge is the result of "hundreds of years of creative human endeavor" (NRC, 2013, p.9).

The new framework embodies a significant shift from learning about science processes to examining scientific phenomena, and engaging learners in sense-making about the natural world while developing and using science and engineering practices (Berland et al., 2016). The new direction in the framework and embodied in the Next Generation Science Standards (NGSS, Lead States, 2013), attempts to bring about changes in teaching and learning in K-12 classrooms by steering students away from rote learning of scientific processes and toward meaningful science understanding (Berland et al., 2016; Cobb & Jackson, 2011; NRC, 2015). As students examine phenomena and generate evidence-based claims, explanation and reasoning, they experience the wonder of inquiring into the scientific enterprise. It is this engagement in learning that will allow for a deep, conceptual understanding of the nature of science. The overarching goal of the framework for K-12 science education, while building on earlier reforms and the first generation of standards seeks to

> ensure that by the end of 12th grade, all students have some appreciation of the beauty and wonder of science, possess sufficient knowledge of science and engineering to engage in public discussion on related issues, are careful consumers of scientific and technological information related to their everyday lives, are able to continue to learn about science outside of schools and have the skills to enter careers of their choice, including (but not limited to) careers in science, engineering, and technology (NRC, 2012, p.1).

The reform document calls for coherence and learning progression in the development of science across the K-12 continuum. This development occurs in a manner that is consistent with specific characteristics of science and how learning occurs. Furthermore, in learning progression, the sequence of learning builds cumulatively and in developmentally informed ways through K-12. That is, as students progress through the grades, they experience the hierarchical development of the science content knowledge and skills and each of the previous level becomes the foundation for later learning. Drawing extensively on research from the learning sciences, cognitive psychology, and education, the framework notes that "children continually build on and revise their knowledge and abilities, starting from their curiosity about what they see around them and their initial conceptions about how the world works" (NRC, 2012, p.11). The effort to streamline the learning progression of science across the K-12 years, therefore leverages students' prior knowledge and experiences to foster the gradual understanding of challenging science concepts. During instruction, the focus on surfacing prior knowledge gives credence to the understanding, explanations, and sensibilities about the world as constructed by students' lived experiences. Then, over multiple years of schooling and with appropriate support, learning progresses toward a more scientifically based and coherent view of the nature of science.

2.2 Science Education in the Era of STEM

In 2001, the acronym STEM was introduced to the community by the National Science Foundation (NSF) in reference to career fields in the critical disciplines of science, technology, engineering, and mathematics (STEM). Since then, STEM has become the centerpiece of an educational system with a heavy focus on two key issues – STEM education and STEM workforce (Brown, Brown, Reardon, & Merrill, 2011; McComas, 2014). In this rapidly changing world, a strong foundation in STEM education has been touted as a necessity for all young people and is "considered paramount to America's economic competitiveness" (Hansen & Gonzalez, 2014, p. 139). However, highly influenced by science and technology, the accompanying complexities and challenges require lifelong learning skills including the ability to solve difficult problems, and contribute to decisions and solutions to improve the quality of life (Sanders, 2009; Tanenbaum, 2016). STEM education, if taught through real-world applications, can instill a lifelong passion for inquiry and discovery and the development of skills and dispositions relevant for the twenty-first century (Cannady, Greenwald, & Harris, 2014). Even though there have been fits and starts in consistency of approach and its effective enactment, the STEM fields have been positioned as the "gateway to America's continued economic competitiveness and national security, and the price of admissions to higher education and higher standards of living for the country's historically underrepresented populations" (Lewis, Miller, Piché, & Yu, 2015, p.2).

Historically, a description of science included an array of disciplines, some of which included biology, chemistry, physics and earth and space sciences. And, with its intellectual pursuit and high social status, science has a well-established history of integration with mathematics and a relationship with technology that impacts application and problem solving used to contextualize science learning (Becker & Park, 2011). Scientists for centuries have used technological tools to conduct experiments and mathematics and statistics to interpret the data produced by those experiments (Lee, Chauvot, Plankis, Vowell, & Culpepper, 2011; Stinson, Harkness, Meyer, & Stallworth, 2009; Weinberg & Sample McMeeking, 2017). The incorporation of both technology and engineering in support of science learning is therefore not new as indicated in Science for All Americans (Rutherford & Ahlgren, 1990), Benchmarks for Science Literacy (AAAS, 1993), and the National Science Education Standards (NRC, 1996b). However, according to Bybee (2010), even though both technology and engineering were identified in reform documents, because of the low occurrences in school science across the nation, their translation into school programs and instructional practices were not fully realized.

In the current reform efforts in science education, the framework explicitly incorporates engineering practices as an important element in the three-dimensional approach to science learning. Science is therefore, not just a body of knowledge that reflects current understanding of the world but includes a set of practices in the inquiry process that establishes, extends, and refines knowledge (Guzey, Moore, Harwell, & Moreno, 2016; NRC, 2012). The framework provides guidance to the NGSS and sets the precedence for the development of a relevant and appropriate twenty-first century science education. The expectations of what students should know and be able to do are thus translated in standards documents in which science and engineering practices are positioned in a central role. The framework (NRC, 2012) in its influence on the current thinking and practices about science teaching and learning has prioritized engineering and scientific practices while identifying the "coherent set of ideas that can provide a foundation for further thoughts and elaboration of the discipline" (Passmore, 2014). Some scholars suggest that this attention to the engineering practices signals the importance of integration or an interdisciplinary approach within the fields of science, technology, engineering, and mathematics (STEM). Science and engineering practices as essential elements of this new vision of teaching and learning science are therefore more apt to support meaningful understanding of crosscutting concepts and disciplinary core science ideas (Brophy, Klein, Portsmore, & Rogers, 2008; Kőycű & de Vries, 2016; Lachapelle & Cunningham, 2014), and furthering students' understanding of the nature of science.

2.3 Science Teaching and Learning in the Twenty-First Century

The ensuing debates around "higher-quality" science education have in most cases resulted in changes to the definition of rigor and relevance of curriculum. Decisions around science teaching and learning in general have been a point of great deliberations in the public, political and educational domains. Internationally, the attention given to the establishment of content standards and national curriculum has resulted in the emergence of national entities with responsibilities for curriculum policy and development (Schmidt, Wang, & McKnight, 2005). In Australia and in England, the development of the national curriculum was impacted by continued waves of changes in educational governance (Briant & Doherty, 2012). The impending changes attracted vociferous debates around whose best interest should drive the reform efforts and how best to converge the varying agendas. In Australia, high stakes discussions around reform and the development of its first national curriculum referenced imperatives related to national competitiveness, globalization, and the interests of the stakeholders (Briant & Doherty, 2012). In England, discussions of the revision of its long-standing national curriculum included issues of relevance and the essential knowledge and skills at the primary and secondary school levels. Nevertheless, in both countries the process of developing and revising the national curriculum included deliberations among the suite of convergent agendas. Interestingly, in England, the appointed experts and stakeholders were charged to "emulate the world's most successful school systems and combine the international best practices with practices from schools in England" (Department of Education, 2014).

In the U.S., influenced by the decentralized education systems and suspicion of federal control, matters of national curriculum have been relegated to being a political issue (Schmidt et al., 2005). Nevertheless, the Framework for K-12 Science education (NRC, 2012), provided research-based guidance that informed the development of the latest Science Standards – the Next Generation Science Standards (NGSS). The NGSS as the curricular guide includes expectations for what students should learn and be able to do throughout K-12. As these expectations become translated into curriculum, they form the basis for classroom instructions and signal a departure from traditional science teaching and learning. Developments in science teaching are also now heavily influenced by contemporary beliefs about learning garnered from the fields of learning sciences. Collective studies in these research domains have contributed to an increase in our understanding of the principles of knowledge organization and how learning occurs. These studies have provided evidence-based direction to the nature and approach to an effective curriculum and its accompanying instructional activities.

An effective K-12 science program allows students to build understanding of science content knowledge and develop scientific and engineering skills. Students are immersed in practices that include examination of science phenomena, collection and representation of data, generation of claims and reason with oneself and others (Duschl, Schweingruber, & Shouse, 2007). Furthermore, in these instructional activities, students develop and use skills such as critical thinking, communication, and problem-solving. Skills that are important for learning through K-12 educational system into their professional lives and as functional citizens. Although many approaches to science teaching exist, curriculum embracing NGSS require a move away from traditional teaching and learning strategies. What is required is a strategy that allows students to develop an understanding of the nature of science and how science knowledge develops within the context of the 3D approach to science learning proposed in the reform documents. Research shows that to enhance students' conceptual understanding and their abilities to think and communicate more scientifically, teachers' instructional strategies should allow students to engage in inquiry practices as they analyze evidence and justify their claims (Harris et al., 2015; McNeill & Krajcik, 2011).

History records many different reconsiderations and proliferation of meanings associated with inquiry and inquiry-based teaching. Several trends in science education have overtime contributed to the altered models and roles of inquiry. Researchers have identified the following roles, science for scientists approach, science for all, and scientific knowledge as indispensable for participation in the workplace and maintaining global economic (DeBoer, 1991; Duschl & Grandy, 2008). The conceptual and methodological developments have fundamentally changed our beliefs about what it means to do science, engage in scientific inquiry, and the description of science as way of knowing (Brickhouse, 2008). In some areas, scholars have argued for multiple models of inquiry informed by the purpose or the goal of the activity. Regardless, however, much credence is given to inquiry-based science teaching because it allows students to learn science consistent with how scientists do science. Students examine phenomena, argue for the explanations they construct, defend their interpretations of the associated data, and advocate for the designs they propose. In the process, the students develop an understanding of the nature of science and the dynamic processes in knowledge construction.

One of the hallmarks of learning science by doing is engagement in the scientific enterprise and the generation of scientific knowledge. The recognition of argumentation as an epistemic practice associated with the appropriation of scientific discourse (Bricker & Bell, 2008; Duschl et al., 2007), has been given much credence in K-12 science. The focus reflects a prioritization of higher-order thinking skills, scientific and engineering practices, and a departure from the traditional view of science learning as primarily or solely the transfer of a body of knowledge (Osborne et al., 2016). Scientific argumentation as a complex form of reasoning requires domain-specific knowledge and evidence to critique claims (Osborne, Erduran, & Simon, 2004), critical to the production and evaluation and therefore the advancement of knowledge. In the process, students engage with the social construction of scientific ideas as well as learn about the scientific enterprise as they persuade their peers of the validity of their construction (Bricker & Bell, 2008; Duschl & Grandy, 2008; Osborne et al., 2004). Furthermore, students perform language-intensive practices while doing, talking, and writing science (Lee, Quinn, & Valdés, 2013), leading to a deeper understanding of the phenomenon under investigation.

2.3.1 Windows into Science Classrooms

For many years, framed around the progressive principles of John Dewey, educators have advocated for learner centered classrooms, learning by doing, and the enactment of relevant and developmentally appropriate curricular activities. Significantly, all of these factors seem to focus on the position of the learners and in general, how learning is afforded during inquiry-based science teaching. While learning is not determined by teaching, there is a high level of dependency on the processes (Davis, Sumatra, & Luce-Kapler, 2008). That is, the role of the teacher in advancing students' development and achieving society's aspirations is important in the dynamic interplay between teaching and learning.

Effective science teaching is more than the simple engagement of learners. The complexity of the act of teaching requires teachers to apply knowledge from multiple domains in order to facilitate students' learning (Park & Chen, 2012). The conceptions of the knowledge base for teachers have increased in complexity and sophistication and have fostered much research in the area of pedagogical content knowledge (PCK). However, the task of the science teacher still remains the integration of the science content matter, science specific pedagogy and context as needed to create learning opportunities. Involved in this integrative process, are teachers' actions that are usually responsive to how children learn science, what makes the learning of specific topics easy or difficult and utilizing strategies that lead students to an understanding of crosscutting concepts and disciplinary core ideas (NRC, 2012).

Identifying the relationship among teachers' knowledge, their practices, and students' science learning has been a persistent and long-standing focus in science education (Abd-El-Khalick et al., 1998). A deep-rooted belief is that teachers' beliefs about students' learning and teaching can impact the implementation of reform-based curriculum (Brickhouse, 1990, Tobin & McRobbie, 1996). While educational policies, such as new standards like the NGSS, attempt to bring about changes in teaching and learning in K-12 classrooms (Cobb & Jackson, 2011), it is the day-to-day practices of teachers that exert the most powerful influence on learning (Clotfelter, Ladd, & Vigdor, 2007; Windschitl, Thompson, Braaten, & Stroupe, 2012).

Fundamental to the achievement of the goals of reform, is the need for a cadre of science teachers, steeped in effective science teaching and learning consistent with the current reform efforts. In order to implement the vision, all science teachers currently in the system will need some level of continued professional development. Many science teachers may not have the appropriate expertise to meet the expectations in the reform. In addition, many teachers report that they have never experienced teaching or learning science as inquiry (Windschitl, 2003), and their traditional socialization into school science is inconsistent with the current reform. To realize the reform, science teachers will need to adjust their instructions, seamlessly integrating the three dimensions of science learning (Bismack, Arias, Davis, & Palincsar, 2015; Harris et al., 2015; Mesa, Pringle, & King, 2014), and engage their students daily in learning experiences focused on exploring and understanding scientific phenomena.

To meet the learning needs of all learners, teachers will need to develop a body of knowledge that includes content knowledge, science-specific pedagogy, and an understanding of how children learn science. The realization of transformed teaching practices and increase in K-12 science achievement have implications for the nature of PD for science teachers; a PD experience that prepares, positions, and facilitates teachers' development as reflective practitioners. Such empowerment of the teaching profession has the potential to be responsive therefore delaying the need for the next mayor wave of reform in science education.

References

AAAS (American Association for the Advancement of Science). (1993). *Benchmarks for science literacy*. New York, NY: Oxford University Press.

Abd-El-Khalick, F., Bell, R. L., & Lederman, N. G. (1998). The nature of science and instructional practice: Making the unnatural natural. *Science Education, 82*(4), 417–436.

Becker, K., & Park, K. (2011). Effects of integrative approaches among science, technology, engineering, and mathematics (STEM) subjects on students' learning: A preliminary meta-analysis. *Journal of STEM Education: Innovations & Research, 12*, 23–36.

Berland, L. K., Schwarz, C. V., Krist, C., Kenyon, L., Lo, A. S., & Reiser, B. J. (2016). Epistemologies in practice: Making scientific practices meaningful for students. *Journal of Research in Science Teaching, 53*(7), 1082–1112.

Bismack, A. A., Arias, A. M., Davis, E. A., & Palincsar, A. S. (2015). Examining student work for\ evidence of teacher uptake of educative curriculum materials. *Journal of Research in Science Teaching, 52*(6), 816–846.

Bransford, J., Brown, A., & Cocking, R. (Eds.). (1999). How people learn. Washington, DC: National Academy Press.

Briant, E., & Doherty, C. (2012). Teacher educators mediating curricular reform: anticipating the Australian curriculum. *Teaching Education, 23*(1), 51–69. https://doi.org/10.1080/10476210.2011.6206

Bricker, L. A., & Bell, P. (2008). Conceptualizations of argumentation from science studies and the learning sciences and their implications for the practices of science education. *Science Education, 92*(3), 473–498.

Brickhouse, N. (2008). What is inquiry? To whom should it be authentic in. In R. A. Duschl, R. E. Grandy, & R. E (Eds.), *Teaching scientific inquiry: Recommendations for research and implementation* (pp. 284–303). Rotterdam, The Netherlands: Sense Publishers.

Brickhouse, N. W. (1990). Teachers' beliefs about the nature of science and their relationship to classroom practice. *Journal of Teacher Education, 41*, 53–62.

Brophy, S., Klein, S., Portsmore, M., & Rogers, C. (2008). Advancing engineering education in P-12 classrooms. *Journal of Engineering Education, 97*(3), 369–387.

Brown, R., Brown, J., Reardon, K., & Merrill, C. (2011). Understanding STEM: current perceptions. *Technology and Engineering Teacher, 70*(6), 5–9.

Bybee, R. W. (1993). *Reforming science education. Social perspectives & personal reflections*. New York, NY: Teachers College Press.

Bybee, R. W. (2010). Advancing STEM education: A 2020 vision. *Technology and engineering teacher, 70*(1), 30–35.

Cannady, M. A., Greenwald, E., & Harris, K. N. (2014). Problematizing the STEM pipeline metaphor: is the STEM pipeline metaphor serving our students and the STEM workforce? *Science Education, 98*(3), 443–460.

Clotfelter, C. T., Ladd, H. F., & Vigdor, J. L. (2007). Teacher credentials and student achievement: Longitudinal analysis with student fixed effects. *Economics of Education Review, 26*(6), 673–682.

Cobb, P., & Jackson, K. (2011). Towards an empirically grounded theory of action for improving the quality of mathematics teaching at scale. *Mathematics Teacher Education and Development, 13*(1), 6–33.

Davis, B., Sumatra, D., & Luce-Kapler, R. (2008). *Engaging minds: Changing teaching in complex ways*. New York, NY: Taylor & Francis.

DeBoer, G. E. (1991). *A history of ideas in science education: Implications for practice*. New York, NY: Teachers College Press.

Department of Education (2014). The National Curriculum for England. Retrieved: https://www.gov.uk/government/collections/national-curriculum

Donovan, S., & Bransford, J. (2005). *How students learn: History, mathematics, and science in the classroom*. Washington, DC: The National Academies Press. https://doi.org/10.17226/10126

Duschl, R. A., & Grandy, R. E. (Eds.). (2008). *Teaching scientific inquiry: Recommendations for research and implementation*. Rotterdam, Netherlands: Sense Publishers.

Duschl, R. A., Schweingruber, H. A., & Shouse, A. W. (Eds.). (2007). *Taking science to school: Learning and teaching science in grades K-8*. Washington, DC: National Academies Press.

Guzey, S. S., Moore, T. J., Harwell, M., & Moreno, M. (2016). STEM integration in middle school life science: Student learning and attitudes. *Journal of Science Education and Technology, 25*(4), 550–560.

Hansen, M., & Gonzalez, T. (2014). Investigating the relationship between STEM learning principles and student achievement in math and science. *American Journal of Education, 120*(2), 139–171.

Harris, C. J., Penuel, W. R., D'Angelo, C. M., DeBarger, A. H., Gallagher, L. P., Kennedy, C. A., & Krajcik, J. S. (2015). Impact of project-based curriculum materials on student learning in science: Results of a randomized controlled trial. *Journal of Research in Science Teaching, 52*(10), 1362–1385.

Kőycű, Ü., & de Vries, M. J. (2016). What preconceptions and attitudes about engineering are prevalent amongst upper secondary school pupils? An international study. *International Journal of Technology and Design Education, 26*(2), 243–258.

Lachapelle, C. P., & Cunningham, C. M. (2014). Engineering in elementary schools. In *Engineering in pre-college settings: Synthesizing research, policy, and practices* (pp. 61–88). West Lafayette, IN: Purdue University Press.

Lee, M. M., Chauvot, J., Plankis, B., Vowell, J., & Culpepper, S. (2011). Integrating to learn and learning to integrate: A case study of an online master's program on science–mathematics integration for middle school teachers. *The Internet and Higher Education, 14*(3), 191–200.

Lee, O., Quinn, H., & Valdés, G. (2013). Science and language for English language learners in relation to Next Generation Science Standards and with implications for Common Core State Standards for English language arts and mathematics. *Educational Researcher, 42*(4), 223–233.

Lehrer, R., & Schauble, L. (2006). Scientific thinking and science literacy. In K. A. Renninger, I. E. Sigel, W. Damon, & R. M. Lerner (Eds.), *Handbook of child psychology: Child psychology in practice* (pp. 153–196). Hoboken, NJ: Wiley.

Lewis, T., Miller, J., Piché, D., & Yu, C. (2015). Advancing Equity through More and Better STEM Learning. The Leadership Conference.

McComas, W. F. (2014). STEM: Science, technology, engineering, and mathematics. In *The language of science education* (pp. 102–103). Rotterdam, The Netherland: Sense Publishers.

McNeill, K. L., & Krajcik, J. S. (2011). *Supporting grade 5–8 students in constructing explanations in science: the claim, evidence, and reasoning framework for talk and writing*. London, UK: Pearson.

Mesa, J. C., Pringle, R. M., & King, N. (2014). Surfacing students' prior knowledge in middle school science classrooms: Exception or the Rule? *Middle Grades Research Journal, 9*, 3.

National Academies of Sciences, Engineering, and Medicine. (2007). *Rising above the gathering storm: Energizing and employing America for a brighter economic future*. Washington, DC: The National Academies Press. https://doi.org/10.17226/11463

National Commission on Excellence in Education. (1983). A nation at risk: the imperative for educational reform: A Report to the Nation and the Secretary of Education United States Department of Education. https://www.edreform.com/wp-content/uploads/2013/02/A_Nation_At_Risk_1983.pdf

National Research Council. (1996a). *National science education standards*. Washington, DC: National Academies Press.

National Research Council. (1996b). *National science education standards: Observe interact, change, learn*. Washington, DC: National Academy Press.

National Research Council. (2007). *Taking science to school: Learning and teaching science in grades K-8*. Washington, DC: National Academies Press.

National Research Council. (2012). *A framework for K-12 science education: Practices, crosscutting concepts, and core ideas*. Washington, DC: National Academies Press.

National Research Council. (2015). *Guide to implementing the next generation science standards*. Washington, DC: The National Academies Press.

NGSS Lead States. (2013). *Next Generation Science Standards: For states, by states*. Washington, DC: The National Academies Press.

Osborne, J., Erduran, S., & Simon, S. (2004). Enhancing the quality of argumentation in school science. *Journal of Research in Science Teaching, 41*(10), 994–1020.

Osborne, J. F., Henderson, J. B., MacPherson, A., Szu, E., Wild, A., & Yao, S. Y. (2016). The development and validation of a learning progression for argumentation in science. *Journal of Research in Science Teaching, 53*(6), 821–846.

Park, S., & Chen, Y. C. (2012). Mapping out the integration of the components of pedagogical content knowledge (PCK): Examples from high school biology classrooms. *Journal of Research in Science Teaching, 49*(7), 922–941.

Passmore, C. (2014). Implementing the Next Generation Science Standards: How your classroom is framed is as important as what you do in it. *NSTA Blog, 11*, 10–14.

Rudolph, J. L. (2002). *Scientists in the classroom: The cold war reconstruction of American science education*. New York, NY: Palgrave.

Rutherford, J., & Ahlgren, A. (1990). Science for All Americans. American Association for the Advancement of Science Projects 2061.

Sanders, M. (2009). Integrative STEM education: primer. *The Technology Teacher, 68*(4), 20–26.

Schmidt, W. H., Burroughs, N. A., & Cogan, L. S. (2013). On the road to reform: K-12 science education in the United States. *Bridges, 43*(1), 7–14.

Schmidt, W. H., Wang, H. A., & McKnight, C. C. (2005). Curriculum coherence: An examination of US mathematics and science content standards from an international perspective. *Journal of Curriculum Studies, 37*(5), 525–559.

Stine, D. D. (2008). *The Manhattan Project, the Apollo program, and federal energy technology R&D programs: A comparative analysis*. Washington, DC: Congressional Research Service, Library of Congress.

Stinson, K., Harkness, S. S., Meyer, H., & Stallworth, J. (2009). Mathematics and science integration: Models and characterizations. *School Science and Mathematics, 109*(3), 153–161.

Tanenbaum, C. (2016). *STEM 2026: A vision for innovation in STEM education*. Washington, DC: US Department of Education.

Timms, M., Moyle, K., Weldon, P., & Mitchell, P. (2018). *Challenges in Stem Learning in Australian Schools: Literature and Policy Review*. Melbourne, VIC: Australian Council For Educational Research.

Tobin, K., & McRobbie, C. J. (1996). Cultural myths as constraints to the enacted science curriculum. *Science Education, 80*(2), 223–241.

Weinberg, A. E., & Sample McMeeking, L. B. (2017). Toward meaningful interdisciplinary education: High school teachers' views of mathematics and science integration. *School Science and Mathematics, 117*(5), 204–213.

Windschitl, M. (2003). Inquiry projects in science teacher education: What can investigative experiences reveal about teacher thinking and eventual classroom practice? *Science Education, 87*(1), 112–143.

Windschitl, M., Thompson, J., Braaten, M., & Stroupe, D. (2012). Proposing a core set of instructional practices and tools for teachers of science. *Science Education, 96*(5), 878–903.

Chapter 3
University of Florida Unites Teachers to Reform Education in Science (U-FUTuRES): A Reform-Based and Comprehensive Professional Development Program

The publication of A framework for K-12 Science Education: Practices, Crosscutting Concepts, and Core Ideas (National Research Council, 2012) signaled the beginning of the current reform in science education. A reform in which high value is placed on the role of science and engineering practices in the construction of science content knowledge and the embrace of a more authentic perspective of science proficiency for K-12 learners (Knight-Bardsley, & McNeill, 2016; Lewis, Baker, & Helding, 2015). The framework proposes a K-12 science education in which students engage in scientific and engineering practices and apply crosscutting concepts to deepen their understanding of core ideas (National Research Council, 2012). The successful implementation of the new standards and the curriculum developed to achieve the changes, is dependent on teachers' preparedness. If teachers are to translate the standards into reform-based teaching, they will need to develop meaningful and appropriate levels of the disciplinary content knowledge and be prepared to enact practices informed by current research (Bismack, Arias, Davis, & Palincsar, 2015; Wilson, Schweingruber, & Nielsen, 2015).

While the elements of the new reform will be incorporated into preservice teacher education, inservice teachers will require both formal and informal learning experiences to be able to transform the conceptions of the reform into practices. There is, however, an abundance of literature lauding the importance of professional development (PD) programs in bringing about substantive changes in instructional practices and ultimately student science achievement (Borko, Jacobs, & Koeliner, 2010; Feiman-Nemser, 2001; Guskey, 2000; Paik et al., 2011; Wilson, 2013). This chapter describes the structure and process of a research-based comprehensive PD program designed to transform science teaching in the middle grades. The model included some of the general research-based elements of high-quality PD identified in the literature along with discipline-specific features recommended by science educators. The PD program, University of Florida Unites Teachers to Reform Education in Science (U-FUTuRES), was developed as a partnership in response to the need for quality science teaching in partnering school districts.

R. M. Pringle, *Researching Practitioner Inquiry as Professional Development*,
https://doi.org/10.1007/978-3-030-59550-0_3

3.1 The Context

The current reform in science teaching has generated new-levels of expectations for middle school teachers. Teachers now need a high level of science proficiencies to effectively engage students in scientific and engineering practices and deepen their understanding of core ideas. Twenty-first century teaching also requires teachers to have skills to prepare students as critical thinkers and autonomous learners while attending to science learning in diverse classrooms. In the initial interviews conducted for selecting participants for the U-FUTuRES PD program, the middle school teachers were asked to share their experiences as science learners and the professional preparation for their assignment as science teachers. While the responses were varied, in general, they expressed levels of unpreparedness to teach in a manner consistent with reforms. One teacher noted,

> When I was looking for a job, they had a science teacher position available. They offered it to me, actually, I applied for an English teacher position. The administrator said I was to give it a try, we will work with you. And I fell in love with that. It's a lot of fun teaching science. I think it's probably more fun than I ever expected. You know of course in my first years I had to learn with the kids. I had to teach all three levels at years sixth, seventh and eighth. And I learned to be very creative. We learned at the same time. (Teacher Interview, 2011)

Another teacher who was first employed as the school's football coach was later assigned to teach physical science to 8th graders. He expressed his love for science but explained:

> The textbook was my main source for teaching my students. I had very good kids who wanted to play football so they were good at completing their science work. It was a struggle for me though coming from a sales position to teaching 8th grade science. I remember my 8th grade science project but struggled with teaching that at my school because I was not trained for inquiry and the school did not provide the equipment that we would need. (Teacher Interview, 2013)

Only a small percentage of the teachers enrolled in the PD program were science or science education majors. The science education majors matriculated through teacher education programs and had experiences with inquiry-based science teaching. Others of the teachers recommended for enrollment in our PD program completed their formal science coursework at their community colleges. And, for others, their science course selections were intended to satisfy the requirements of their general education program of their degrees. These courses were certainly not intended to prepare them for teaching science in middle schools. The extent of their science knowledge was mostly garnered over time from teaching at the assigned grade level as shared by one of the teachers:

> I really came to like the earth science over the years. I'm used to it now. I feel like I had so much experience learning with the students. Now, I can really impact them with my knowledge and things that I have tried and learned from Discovery education. So, you know, it's like trial and error (Teacher Interview, 2013).

The extent of the teachers' knowledge was limited to the science content contained in their district adopted science texts (Pringle, Mesa, & Hayes, 2018), and which became their curriculum and their reference for the content knowledge. This is certainly inadequate preparation to support effective teaching of science as envisioned in the reform. The middle grades are situated at a critical juncture in the K-12 trajectory. Here, teaching and learning experiences serve to foster the proficiencies and dispositions required for high school and post-secondary education. Another key point, researchers concerned with adolescents, posit that students in the middle grades are vulnerable to failing in school and losing interest in science (Schriver & Czerniak, 1999). While the teachers in the PD program represent a subset of the population of middle school teaching force, educators agree there is much room for improvement in teaching and learning science at this level of the education system.

3.2 Professional Development in the Service of Science Reforms

The new approach to science teaching and learning urged on by A Framework for K-12 Science Education: Practices, Crosscutting Concepts, and Core Ideas (NRC, 2012) and encapsulated in the national science standards (NGSS, Lead States, 2013) has led to the need for significant investment and efforts in enhancing teachers' learning. Reform documents have clearly recognized teachers' PD as the linchpin for the implementation of reform in education and also as a viable means to support practicing teachers in their professional learning (NRC, 2012). Since the release of the new framework, many efforts have focused on designing and developing PD to build teachers' instructional skills and science content knowledge. PD programs, described as systematic efforts to increasing student learning, have the potential to achieve such by transforming teacher practice, beliefs and attitudes (Guskey, 2002). Furthermore, if well-designed and executed, PD can be supportive of school reforms initiated from within the learning environments and then moving outwards (Elmore, 2004).

PD has a fairly long history in its development and impact on in-service teachers' professional learning. Teachers who engage in robust professional learning activities are more likely to improve their teaching practice and student learning (Borko, 2004; Johnson, Kraft, & Papay, 2012; Loucks-Horsley, Stiles, Mundry, Love, & Hewson, 2010). Research conducted over many years has provided a rich body of knowledge on "best practices" in supporting teachers' learning during PD. In fact, many models have been developed, and overtime, a collective understanding of core and structural features of effectiveness have emerged (Desimone, 2009; Garet, Porter, Desimone, Birman, & Yoon, 2001; Lee & Buxton, 2013; Penuel, Fishman, Yamaguchi, & Gallagher, 2007).

The well-established body of work has identified key aspects of high-quality PD, deemed most likely to result in deepening teachers' knowledge and changing teaching practices. These general and critical aspects of effective PD for science teachers include a clear focus on classroom practices to increase students' achievement. Effective PDs therefore have as their objectives, the deepening of subject matter and pedagogical knowledge; engaging teachers as active learners within a collaborative context such as a community of learners; enacting coherent program activities aligned with school policies, practices and expectations; substantive duration and sustainability both in intensity and contact hours; and facilitate systematic reflection (Bellanca & Brandt, 2010; Borko, 2004; Darling-Hammond and Bransford, 2005; Garet et al., 2001; Lee & Buxton, 2013; Lotter, Rushton, & Singer, 2013; Loucks-Horsley et al., 2010; van Driel, Meirink, Van Veen, & Zwart, 2012). Others suggest that PD activities should also include the support of teachers in their endeavor to embed formative assessment during instruction which allows for close consideration of student reasoning and identification of issues related to content knowledge and teaching strategy (Black, 1993; Borko et al., 2010; Shepard, 2005). The agreement among scholars is that when teachers engage in formative assessment, they elicit and interpret their students' ideas, examine student work, and use the students' current understanding to inform their instructional decisions and actions (Black & William, 2009; Borko et al., 2010; Dixon, & Williams, 2003; Shepard, 2005).

Researchers examining the link between PD for science teachers and learning, posited that the effects on student achievement are mediated by changes in teacher knowledge and skills as well as changes in instructional practice (Lakshmanan, Heath, Perlmutter, & Elder, 2011; Yoon, Duncan, Lee, Scarlos, & Shapley, 2007). To realize the reform ideals, science teachers must learn how to adjust their instruction to seamlessly integrate the three dimensions of science learning (Bismack et al., 2015; Harris et al., 2015; Mesa, Pringle, & King, 2014), and engage their students daily in learning experiences focused on exploring and understanding scientific phenomena rather than just learning about them. Teachers also need to learn how to link students' lived and cultural experiences to curriculum goals in order to make instruction relevant (Darling-Hammond, & Bransford, 2005; Ladson-Billings, 1994). When teachers are unaware of students' interests and life experiences, they fail to engage with the students' prior knowledge, thus limiting the process of learning.

Our PD embraces the notion that teachers' learning and improved teaching are central to translating reform efforts into practice at the school level. The PD, with a focus on students' learning, was designed to build teachers' content knowledge, immerse teachers in reform-based teaching practices within collaborative communities of learners, and develop the mechanism for long-term support beyond the PD program (Darling-Hammond, Wei, Andree, Richardson, & Orphanos, 2009; Knapp, 2003; Loucks-Horsley et al., 2010; Wilson, 2013). Going further, science educators engaged in PD work have declared that teachers, as learners, should be engaged in inquiry-based science and with opportunities to develop, practice, reflect, and refine their practices within supportive communities (Lotter et al., 2013; Loucks-Horsley et al., 2010; Whitworth & Chiu, 2015). These principles guided the development of the PD program – *The University of Florida Unites Teachers to Reform Education in Science project (U-FUTuRES)* with the aim of preparing and supporting science teachers in partnering school districts to effectively implement the inquiry-based curriculum, Investigating and Questioning our World through Science and Technology (IQWST). Specifically, the PD program engaged the middle school science teachers in experiences related to science learning, teaching, assessment, and reflection on practice, as they made connections among their learning, reform-based teaching practices, and student learning.

3.3 Theoretical Framework

The development and enactment of U-FUTURES was guided by Guskey's (1989) tested theory of teacher learning and constructivism as an epistemology, which has major influence in contemporary science education (Pelech & Pieper, 2010; Tobin, 1993). Guskey's theory highlights the process of teacher change and the influence of their learning on their own practices. While educators suggest that the process of changing teaching practices occurs over time (Hewson, 2007), there is a deliberate relationship between teacher change and learning that is developmentally and experientially based. Guskey (2002), in discussing ways to promote meaningful changes in science teaching, posited, that reform-based teaching practices are reinforced and sustained when teachers observe notable shifts in student engagement and learning outcomes.

In science PD, teachers as learners construct new knowledge of science ideas and practices, an understanding of instructional strategies consistent with current vision, and the skills to implement those strategies in their classrooms. Thus, constructivism as a theoretical framework guides our work as PD providers and designers of teacher education programs. Constructivism as an epistemology views learning as an active and subjective process that is shaped and structured by the learners' experiences (Tobin, 1993). Furthermore, new knowledge is generated through intellectual discourse rather than passive reception of information. That is, the process of learning involves the adaptive interaction between existing knowledge or beliefs of the learner and new knowledge and individual experiences. The constructivist view of learning also emphasizes the social component of knowledge and as such considers the social and cultural contexts in which the learners' ideas are generated. This view of the nature of learning places the learners, their experiences, and the newly constructed knowledge at the center of teaching, learning, and the curriculum. Giving credence to the social context of learning including tools and cultural objects and the importance of prior knowledge, Vygotsky (1978) acknowledges that "the socially elaborated contents of human knowledge and the cognitive strategies necessary for their internalization are evoked in the learners according to their actual levels" (p.131).

Constructivism is undoubtedly a major theoretical influence in contemporary science education (Pelech & Pieper, 2010; Tobin, 1993). As a theory, it proposes suggestions for classroom instructions and teachers' PD aimed at facilitating conceptual change and improved student achievement. While constructivism provides an adequate basis for accomplishing the goals of teaching, learning, and curriculum, some researchers object to the prescription of 'best practices' for constructivist teaching. They contend that constructivism must not be considered a set of methods but rather used as a referent for making decisions about practice (Crowther, 1999; Tippins, Tobin & Hoock, 1993; Windschitl, 2002). The design and enactment of our PD was informed by the tenets of constructivism. The model included a sequence of activities that began with surfacing and challenging teachers' previous conceptions and which lead to the construction of viable patterns of teaching practices (Brooks & Brooks, 1999; Pelech &Pieper, 2010). Other activities included cooperative learning, self and peer assessment, and reflective practices.

Teachers, as learners, bring to PD programs, a wide range of knowledge about teaching and learning garnered from their individual practices. In the program, teachers, as a community of learners, were afforded opportunities to collaboratively engage in reflection and discourse with their peers as they supported each other's learning. These collaborative practices support the teachers' continued learning and are strengthened through the strategic development of a *community of learners* – connected via online interactions and encouraged by activities in the annual *Summer Science Summit*. This learner-centered approach with emphasis placed on community discourse is central to our embrace of constructivism.

3.4 The Professional Development Program

3.4.1 Setting the Context

The achievement gap between populations of science learners is proving stubbornly persistent. Documented challenges faced by schools serving rural districts, low socio-economic status (SES), and high minority populations include high teacher turnover and lack of preparedness to teach in assigned subject areas (Duschl, Schweingruber, & Shouse, 2007; Lankford, Loeb, & Wyckoff, 2002; Lewis, Baker, & Jepson, 2000; Ruby, 2006). These challenges have undoubtedly manifested themselves in persistently low performance and widening achievement gaps when students in these areas are compared to their counterparts in affluent school districts. Research in schools with high minority populations and low SES has consistently shown that when science is taught, engaging teaching activities are limited and science-specific pedagogy is rarely observed (Gibson & Chase, 2002; Hewson, Kahle, Scantlebury, & Davies, 2001). Regardless of their cultural and socio-economic status, all middle school students require a solid foundation in science and related skills to become functional adults to reduce the possibility of a lifetime of unemployment.

3.4.2 *The Partnership*

Educational partnerships between universities and public schools are not new (Goodlad, 2004; Zhang, McInerney & Frechtling, 2010) but there is renewed interest in developing strategic coalitions to support school change and improve students' achievement. Several scholars have stressed the importance of developing school-university partnerships as alliances focused on addressing policy and practices central to the issues being faced by schools and districts. Some of these issues are directly related to students' science achievement. But, typically, when Institutes of Higher Education (IHE) and school personnel become engaged in partnership work, the innovations are developed independently of the context to be served (Coburn, Penuel, & Geil, 2013), and not rooted in the needs of the schools. This, according to many researchers is a recipe for failure and could be the reason why much of the work involving schools and IHE personnel become limited in their scope, achievement, and sustainability. The principles of mutualism, commitment to long-term collaboration, and abiding efforts to build and maintain trusting relationships were key principles that guided our deliberation in the early stages of developing the partnership. Also, the priorities of the schools and district administrators became one of the focal points around which the program's activities were developed, enacted, and constantly revised.

The partnership was intentionally organized to be responsive to the needs of school districts as they identified their long and short-term goals to transform science teaching and learning in their middle schools. As seen in Table 3.1, there were four major partners – schools and district personnel, P. K. Yonge Development Research School, scientists from the College of Liberal Arts and Sciences, and science educators from the College of Education.

Schools and District Personnel Local Education Agencies (LEAs) were invited to participate in the project based on their interest and readiness to commit to a district wide effort to transform middle school science. A large percentage of their students qualified for free or reduced lunch and their performance on the state science achievement test consistently lagged behind the state average. In addition to providing administrative and management functions, school and district leaders played significant roles in planning and implementing professional development and in securing opportunities to facilitate teacher change.

IHE Partners – The **scientists from the College of Liberal Arts and Sciences** along with their graduate students were integral to the collaborative and continued enhancement of the teacher education program. The scientists learned about the middle school teaching contexts and became familiar with the requirements of the reform efforts in science education. Beyond the deepening of the science content knowledge, they also modeled best practices as they immersed the teachers in inquiry-based teaching and learning. Morrison and Estes (2007), in discussing the benefits of scientists in PD, noted that their involvement was an effective strategy to support the middle school teachers as they learn the content knowledge. Others

Table 3.1 U-FUTuRES program partners

Partners	Contributions
School and district administrators	Identify specific areas of need as determined by students' science achievement patterns Recommend teacher participants Provide teacher support to offset teacher travel during school hours Provide/supplement tuition where applicable
P.K. Yonge Developmental Research School	Coordinate and maintain the interactions with all stakeholders Assist IHE in understanding current state and district education policies, challenges, and opportunities Provide demonstration site for teachers and school/district leaders to observe reform-based science teaching Science teachers in partnership with IHE to inform course content and activities Assist IHE in connecting with school and district personnel thus expanding the partnership
IHE Partners – Science	Co-develop and teach science courses with support from science education partners from the College of Education Revise science courses in collaboration with input from school personnel and science educators
IHE Partners – College of Education	Facilitate the design and development, and continuous revisions of courses Ensure coherence among instructors and project personnel Coordinate consultation with school district partners and PK Yonge DRS to make needed adaptations to support success of teacher participants

agree, that deepening teachers' understanding leads to an increase in teachers' efficacy as science teachers. The IHE scientists co-developed and taught the science courses with support from science education partners from the College of Education. In addition to the co-development of the specialized science courses and ensuring their relevance and appropriateness for the middle school teachers, the IHE scientists participated in all aspects of the program. The activities included the face-to-face meetings, Summer Science Institute, and pre-post course reflections. Their presence and interactions with the teachers and program administrators created an excitement for science content, thus impacting the attitudes of the middle grade teachers.

IHE Partners – P.K. Yonge Developmental Research School and Science Educators from the College of Education: A notable aspect of developing this partnership was borne out of the ongoing relationship and involvement of one of the key project partners. The director of P. K. Yonge Developmental Research School (PKY) was a contributing member of the Northeast Florida Educational Consortium (NEFEC), a support organization for 15 poor, small and rural school districts. Her involvement in NEFEC afforded her a wealth of partnership experiences, resourcefulness, and networking that were valuable to the PD program. The director coordinated the development of the partnership, maintained the institutional linkages with all stakeholders, and facilitated the IHE in understanding the current education policies and school contexts, challenges, and opportunities. Furthermore, PKY became the demonstration site for teachers and schools and district leaders to observe fully developed reform-based science teaching practices.

The science educators coordinated the activities of the PD program including the design, development, and continuous revisions of all courses in the program, and facilitated instructor preparation to ensure coherence. The science leadership institute was situated in the College of Education. The formal graduate courses were housed and enacted through the university's learning management system within the School of Teaching and Learning. The education courses were developed and taught by the science education faculty and graduate students.

The theory of action is built upon the model that increasing teacher knowledge and skills will lead to improved teaching practices which in turn increases student outcomes. Teachers' willingness to embrace and reinforce the change in their practice is usually determined by the impact on their students (Guskey, 2002). As seen in Fig. 3.1, the model included (a) district and school-level administrators, (b) a specially-designed online and job-embedded science education graduate degree program—the Science Teacher Leadership Institute (STLI), and (c) cadre meetings and summer summits that included complementary learning activities. Each of these components and their activities were selected in accordance with research-based recommendations identified in high-quality PD. The project activities were facilitated in both face-to-face and online interactions.

The school and district leaders, as partners in the program, ascertained the needs of their schools, selected and recommended teacher participants, and provided the

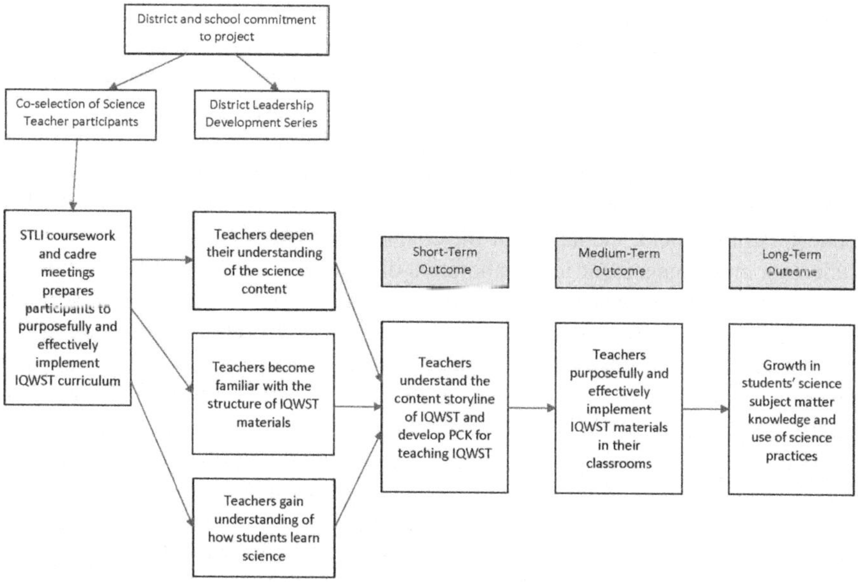

Fig. 3.1 The comprehensive PD model. (Pringle et al., 2018)

necessary support for the teachers to complete the program. Inputs from the schools and districts also informed areas of the program including the development and implementation of a District Leadership Development Series (DLDS). The partnering LEAs also identified a leadership team that worked with project partners during the quarterly DLDS meetings in developing affordable and sustainable approaches to sustain the reform efforts in the districts. In addition, the LEAs as invested partners were integral to the development of long-term plans for transforming science teaching and improving academic outcomes of their students. Their involvements also had the potential to sustain the changes initiated in the PD program and offer support to reduce barriers to the teachers' continued learning (Whitworth & Chiu, 2015).

3.4.3 The Science Teacher Leadership Institute

The science teacher leadership was a two-year masters' program in which the teachers earned a graduate degree (MAE in science education). The program included eight graduate level courses, five of which focused on science-specific teaching and learning and leadership, three science content courses, and culminated with the teachers conducting the inquiry which served as the program's capstone project. Collectively, all the courses shown in Table 3.2 were specifically directed to developing teachers' science content knowledge, science-specific pedagogy, a deepening of the teachers' understanding of the curricula and standards they were expected to teach, and a leadership course that positioned them as agents of change in their schools and districts.

The master's program being online and job-embedded, provided just-in-time opportunities for the teachers to develop and adapt lessons in response to the learning in each of the courses. The timeliness of the ongoing support provided in the program allowed the teachers to connect to their daily guiding of student thinking in science, practice ideals of reform-based teaching and generate questions about effective science teaching and learning. Consistent with constructivism and teacher educators in the field, the nature of the course experiences in facilitating teacher's reflection and engagement in collaborative discourses also allowed for the refinement of their understanding while supporting the learning of their peers (Capps, Crawford, & Constas, 2012; Desimone, Porter, Garet, Yoon, & Birman, 2002; Duschl et al., 2007; Garet et al., 2001; Loucks-Horsley et al., 2010; Luft and Hewson, 2014; Whitworth and Chiu, 2015; Wilson, 2013). The STLI's focus on developing teachers' content knowledge through a combination of inquiry-based lab experiences, readings, on-line investigations, and discussions was also consistent with the renewed emphasis on the importance of subject-matter in designing high-quality professional development. The researchers contend that increasing teachers' content knowledge, was more likely to increase knowledge and skills (Banilower, Boyd, Pasley, & Weiss, 2006; Clary et al., 2018; Garet et al., 2001), thus

Table 3.2 STLI program of study

Courses	Description
Inquiry-based Science Teaching	Focuses on inquiry-based science teaching and approach to K-12 science as put forth by current reform in science education including core ideas, cross-cutting concepts and scientific and engineering practices, and the four strands of science proficiency. The course emphasizes the common base of well-researched knowledge about learning and science-specific teaching strategies, and facilitate the adoption of an inquiry stance
Physics and Chemistry for Middle School Teachers	Provides the foundational knowledge for teaching Chemistry and Physics at the middle school level. Common alternative conceptions specific to introductory Physics and Chemistry are emphasized along with other content-specific pedagogy
Diversity and Equity in Science Teaching	Explores the idea of providing all middle school students with equitable access to an inquiry-based science education, and examine possible barriers to such access. This course allows teachers to develop a variety of inclusive strategies for teaching science to diverse populations in middle schools
Current Topics in Earth and Space Sciences for Teachers	Focuses on understanding the Earth as one interconnected system, in which changes occur over vast scales of time and space, and in which biogeochemical systems react to and recover from disturbances. The course emphasizes the application of knowledge to Florida environments and societal issues of local and global significance
Biological Science for Middle School Teachers	Explores the interrelationships of the natural world; how life is maintained, flow of matter and energy in the ecosystem; and how organisms adjust to changes in the environment, through identification and analysis of environmental problems, both natural and anthropogenic
Science Curriculum Development	Teachers examine state standards, learning goals, and assessment in the curriculum development process and adapt their existing science curriculum to reflect inquiry-based science teaching and the three dimensions of science teaching and learning. The course culminates with the development and enactment of a unit of study relevant to their curriculum and embodies the elements of the PD program *Understanding by Design* (Wiggins & McTighe, 2005) and *How People Learn Science* (Donovan & Bransford, 2005) will provide a shared framework for understanding the IQWST curriculum. Participants will deconstruct the IQWST curriculum and other NSF-developed instructional materials and units to deepen their knowledge of inquiry-based science. Training in the IQWST curriculum will also be provided
Science Teacher Leadership and School Change	Participants examine the dynamics of change within educational contexts by focusing on theory, research, and practice as well as principles for managing change. Coursework will prepare STLI participants for the local challenges they will face as reform leaders and develop specific strategies for nurturing active administrator/ teacher participation in school change
Data Driven Science Instruction	**Data Driven Instruction:** A web-based course supported through monthly STL cadre meetings organized as a professional learning community for STL's; participants will study and apply principles and practices of assessment to inform instructional planning and monitoring equity in science achievement and student engagement. Course assignments include analysis of formative and summative curriculum-based science assessments

ensuring the teaching force required to respond to the current vision of science teaching and learning.

The program was deliberate in its design of the courses and complementary PD activities. The activities were tailored to prepare the teachers to transform their teaching in accordance with the suggestions in the reform documents. The new practices specifically include the active learning of science in which students examine phenomena, generate evidence-based claims, and as scientists, experience the cohesive integration of scientific and engineering practices as they learn crosscutting concepts and disciplinary core ideas (National Research Council, 2012; NGSS Lead States, 2013). The Elements of Effective Science Instruction, essential for promoting conceptual change include (1) motivation, (2) surfacing students' prior knowledge, (3) intellectually engaging with examples/phenomena, (4) using evidence to critique claims, and (5) sense-making (Banilower, Cohen, Pasley, & Weiss, 2010) provided the framework for the design, development, and enactment of the courses. These elements, strategically incorporated into the program and aligned to the current reform efforts in science education also facilitated teachers' learning. Designing and enacting professional learning experiences in this way not only immersed the teachers in practices they are being challenged to implement in their classrooms but allowed them to experience such as learners.

3.4.4 *Face-to-Face Components: Cadre Meetings and Summer Institutes*

The PD was comprehensive, comprising of other activities that complemented the STLI and afforded teachers other learning opportunities. In addition to the formal and carefully designed and orchestrated graduate degree in which theoretical bases were introduced, teachers were engaged in workshop activities in which they experienced cognitive conflict and reconstruction of their knowledge as they reflected and shared with their peers, usually arriving at consensus on best practices. The workshops were mainly conducted in regular face-to-face meetings referred to as cadre meetings and during summer science institutes. Cadre meetings were held each month on the campus of the university and included a variety of activities which built upon topics introduced in the formal courses. The meetings' included complementary experiences that reinforced content knowledge (CK) and pedagogical content knowledge (PCK) topics introduced in the formal courses. For example, Appendix A is an agenda of the first day of a cadre meeting that focused on two components of the elements of effective instruction – surfacing prior ideas and sense making. As shown on the agenda, the three-hour session was dedicated to the science education component and occurred during one of the earlier cadre meeting. The topics were selected because of their importance in the instructional sequence and the direct connection to contemporary beliefs about how learning occurs. Other topics explored in the cadre meetings included, assessment and analysis of student work to inform instruction, cultural competence, differentiated instruction and assessment, and collaborative grouping. Subsequently, each cadre meeting was dedicated to topics that emerged from our observations of the teachers' classrooms, recommendations from the evaluators, or teachers requests that emerged from their own observation during their teaching.

The scientists were also engaged in the cadre meetings. They worked with the teachers to revise and refine their understanding of science concepts that were identified as being problematic. Thus, during the face-to-face interactions, the scientists modeled best practices as they immersed the teachers in inquiry-based teaching. Teachers were therefore learning science in ways they were required to implement the curriculum in their own classrooms. The monthly cadre meetings provided a safe space in which the teachers revealed their vulnerabilities but also felt supported in their learning.

The cadre meetings also provided a forum for teachers to reflect on their experiences and to share ideas about their learning in the program. Over time, the group of teachers evolved into a community of learners (Birman, Desimone, Porter, & Garet, 2000; Garet et al., 2001; Loucks-Horsley et al., 2010). They developed a support system among themselves, shared ideas and spoke openly and frankly about their experiences in the STLI. As a community, they celebrated their successes, identified challenges, and deliberated on possible solutions and strategies to increase their effectiveness. The evolution of such collaboration is widely regarded as a powerful tool to support continued teacher learning (Vescio, Ross, & Adams, 2008), and was evident in the level of interactions occurring outside of the formally planned program activities.

U-FUTuRES Summer Summit The summer summits were multi-days residential workshops held on the campus of the university. In the program's first summer institute, the teachers were introduced to the IQWST curriculum by one of the developers. Teachers were introduced to the unique features of the curriculum and experienced the instructional activities as learners. Subsequent summer institutes incorporated activities that were responsive to issues teachers raised as being problematic. These issues arose either from their formal courses or emerged during their teaching. The issues related to science content knowledge were addressed by the IHE professors and their graduate students. In other activities, teachers worked in teams and completed tasks across and within grade level teams. During these activities the focus was on planning and reinforcing strategies such as meaningful discourse and language and literacy in science learning. The goal of these activities was to facilitate a deeper understanding of the hierarchical structure and content knowledge requirements of the curriculum extending across sixth to eighth grade. Then, as the teachers engaged in reflective discourse they offered support to each other's learning, thus reinforcing the supportive nature of the community.

3.4.5 Investigating and Questioning Our World Through Science and Technology (IQWST) Curriculum

Unlike traditional PD or graduate programs with participants from many different settings, the teachers in this program were from the partnering school districts, all of whom were adopting the Investigating and Questioning Our World through Science and Technology (IQWST) curriculum. Teachers face many barriers that punctuate their resistance to adapting their PD experiences into their teaching practices. Chief among these barriers is the translation of practices learned in PD to their curriculum and classroom instructional activities. Teachers need specific preparation and support to enact reform-based curriculum. This is not easy (Schneider & Krajcik, 2002). The IQWST curriculum was the core around which the PD activities were intentionally developed, to ensure ease of translation into teachers' classroom practices. The job-embeddedness of the program allowed the course assignments to be directly related to the teachers' immediate classroom practices around the curriculum. As teachers enacted their learning, they were challenged to engage in reflective practices thus reinforcing and supporting their learning.

IQWST is a standards-based and carefully sequenced middle grades science curriculum (Krajcik, Reiser, Sutherland, & Fortus, 2012). In each of the three academic years, four units, one each in the disciplines of physics, chemistry, biology, and earth science are taught. One essential feature of the curriculum is the coordination of the science content and science and engineering practices across units within each grade level and their progression across the three years. IQWST is organized so that students' understandings of core ideas (e.g., energy, particle nature of matter) and scientific practices (e.g., designing investigations, models, scientific explanations) are developed progressively (Schwarz et al., 2009), over the middle school years. The three-year science experiences are organized around learning goals and being activity-based, the sixth through eighth grade students learn science by developing and using scientific and engineering practices to examine and understand natural phenomena. The embrace of these practices is consistent with science teaching and learning espoused in the reform documents.

The IQWST curriculum includes a number of teaching strategies to engage students in authentic science practices. For example, the curriculum recognizes scientific argumentation and explanation as ongoing processes of building science knowledge. Hence, students engage in complex science practices in which they collect evidence and propose explanations that drive the practice of argumentation. Students critique their arguments in ongoing discourses that further allow for the development of critical thinking and problem-solving skills. As students ask questions of each other as well as the teacher, they engage in supportive literacy in science learning. Rather than being separate from the science content knowledge, reading, writing, thinking, and doing science are seamlessly integrated into the instructional activities. In addition, a unique feature, the driving question board provides a public display of the anchoring questions guiding each of the lessons. The board also operates as a space for students to post their questions during the enactment of the lesson supporting the process of continued inquiry.

Additionally, the IQWST curriculum includes a number of educative features that support teachers in their enactment of the curriculum (Davis & Krajcik, 2005). Educative curricular materials are designed to support teacher learning as well as their students. Accordingly, they offer support for teachers as they think about the content knowledge and the pedagogical strategies beyond the levels suggested for the students (Ball & Cohen, 1996; Davis et al., 2014; Schneider & Krajcik, 2002). There are six types of educative features in IQWST: teacher background knowledge, teaching strategies, teaching alternatives, common conceptions (of students), prerequisite knowledge (of students), and lesson checkpoints. Specifically, these educative materials provide teachers with suggestions and alternative actions to facilitate pedagogical adjustments and adaptations for the specific needs of their students (Pringle, Mesa, & Hayes, 2017). These opportunities further expand teachers' repertoire of instructional practices while supporting further learning about science teaching (Collopy, 2003; Schneider & Krajcik, 2002).

3.5 Conclusion

The comprehensive U-FUTuRES PD program emerged in response to the need for quality in-service education to prepare teachers to teach science as purported in the new vision of science education. The PD program was a deliberate collaborative effort among stake holders and the development of a partnership that included the middle school teachers, teacher educators, scientists, and school and district administrators. Our efforts have resulted in a partnership intent on improving science achievement during the middle school years by transforming science teaching practices and affording teachers timely and ongoing support. The viability of learning informed by constructivism was embraced throughout the development and enactment of the PD program. The middle grade teachers, as learners were first provided with experiences that surfaced and challenged their prior knowledge about science and science teaching and learning. Then, embedded in their classrooms, the teachers explored the viability of the new knowledge as they transformed their teaching to include the requirements in reform documents, the new standards, and the curriculum.

Evaluation data collected by the project's external evaluators (Horizon Research, Inc., 2015), provided much insights into the effectiveness of the PD program. They concluded that the teachers generally had positive perceptions of the PD program. They reported teachers indicated that the courses were generally relevant to their classroom instruction and were effective in deepening their understanding of the content knowledge. The evaluators also reported an increase in teachers' capacity and strategies to meet the learning needs of the students in their diverse classrooms. Furthermore, the evaluators generated claims from qualitative data analysis that provided much insight into the contextual richness of all areas of the program (Yin,

2016). In particular, in representing the views and perspectives of the teachers, the evaluators noted the efficacy of the network that evolved and served to facilitate continued peer collaboration and support among the teachers.

References

Ball, D. L., & Cohen, D. K. (1996). Reform by the book: What is—or might be—the role of curriculum materials in teacher learning and instructional reform? *Educational Researcher, 25*(9), 6–14.

Banilower, E., Cohen, K., Pasley, J., & Weiss, I. (2010). Effective science instruction: What does research tell us? *Educational Researcher, 33*(8), 3–15. Center on Instruction.

Banilower, E. R., Boyd, S. E., Pasley, J. D., & Weiss, I. R. (2006). *Lessons from a decade of mathematics and science reform: A capstone report for the local systemic change through teacher enhancement initiative*. Horizon Research, Inc. (NJ1).

Bellanca, J., & Brandt, R. (2010). *21st century skills: Rethinking how students learn*. Bloomington, IN: Solution Tree Press.

Birman, B. F., Desimone, L., Porter, A. C., & Garet, M. S. (2000). Designing professional development that works. *Educational Leadership, 57*(8), 28–33.

Bismack, A. A., Arias, A. M., Davis, E. A., & Palincsar, A. S. (2015). Examining student work for evidence of teacher uptake of educative curriculum materials. *Journal of Research in Science Teaching, 52*(6), 816–846.

Black, P., & William, D. (2009). Developing the theory of formative assessment. *Educational Assessment, Evaluation and Accountability (Formerly: Journal of Personnel Evaluation in Education), 21*(1), 5.

Black, P. J. (1993). Formative and summative assessment by teachers. *Studies in Science Education, 21*, 49–97.

Borko, H. (2004). Professional development and teacher learning: Mapping the terrain. *Educational Researcher, 33*(8), 3–15.

Borko, H., Jacobs, J., & Koellner, K. (2010). Contemporary approaches to teacher professional development. *International Encyclopedia of Education, 7*(2), 548–556.

Brooks, J. G., & Brooks, M. G. (1999). *In search of understanding: The case for constructivist classrooms*. Alexandria, VA: ASCD.

Capps, D. K., Crawford, B. A., & Constas, M. A. (2012). A review of empirical literature on inquiry professional development: Alignment with best practices and a critique of the findings. *Journal of Science Teacher Education, 23*(3), 291–318.

Clary, R. M., Elder, A., Dunne, J., Saebo, S., Beard, D., Wax, C., & Tucker, D. L. (2018). Beyond the professional development academy: Teachers' retention of discipline-specific science content knowledge throughout a 3-year mathematics and science partnership. *School Science and Mathematics, 118*(3–4), 75–83.

Coburn, C. E., Penuel, W. R., & Geil, K. E. (2013). Practice Partnerships: A Strategy for Leveraging Research for Educational Improvement in School Districts. William T. Grant Foundation.

Collopy, R. (2003). Curriculum materials as a professional development tool: How a mathematics textbook affected two teachers' learning. *The Elementary School Journal, 103*(3), 287–311.

Crowther, D. (1999). Cooperating with constructivism. *Journal of College Science Teaching, 29*(1), 17–23.

Darling-Hammond, L., & Bransford, J. D. (2005). *Preparing teachers for a changing world: What teachers should learn and be able to do*. San Francisco, CA: Jossey-Bass.

Darling-Hammond, L., Wei, R. C., Andree, A., Richardson, N., & Orphanos, S. (2009). *Professional learning in the learning profession* (p. 12). Washington, DC: National Staff Development Council.

Davis, E., Palincsar, A. S., Arias, A. M., Bismack, A. S., Marulis, L., & Iwashyna, S. (2014). Designing educative curriculum materials: A theoretically and empirically driven process. *Harvard Educational Review, 84*(1), 24–52.

Davis, E. A., & Krajcik, J. S. (2005). Designing educative curriculum materials to promote teacher learning. *Educational Researcher, 34*(3), 3–14.

Desimone, L. M. (2009). Improving impact studies of teachers' professional development: Toward better conceptualizations and measures. *Educational Researcher, 38*(3), 181–199.

Desimone, L. M., Porter, A. C., Garet, M. S., Yoon, K. S., & Birman, B. F. (2002). Effects of professional development on teachers' instruction: Results from a three-year longitudinal study. *Educational Evaluation and Policy Analysis, 24*(2), 81–112.

Dixon, H., & Williams, R. (2003). *Teachers' understandings of formative assessment.* In: A paper presented at the Annual Conference of the British Educational Research Association, University of Leeds, 12–15 September 2001. Retrieved from: http://www.leeds.ac.uk/educol/documents/00002533.htm - 12.1.2019

Donovan, S., & Bransford, J. (2005). *How students learn: History, Mathematics, and Science in the Classroom.* Washington, DC: National. Academies Press.

Duschl, R. A., Schweingruber, H. A., & Shouse, A. W. (Eds.). (2007). *Taking science to school: Learning and teaching science in grades K-8.* Washington, DC: National Academies Press.

Elmore, R. F. (2004). *School reform from the inside out: Policy, practice, and performance.* Cambridge, MA: Harvard Education Press.

Feiman-Nemser, S. (2001). From preparation to practice: Designing a continuum to strengthen and sustain teaching. *Teachers College Record, 103*(6). Teachers College, Columbia University, pp. 1013–1055

Garet, M. S., Porter, A. C., Desimone, L., Birman, B. F., & Yoon, K. S. (2001). What makes professional development effective? Results from a national sample of teachers. *American Educational Research Journal, 38*(4), 915–945.

Gibson, H. L., & Chase, C. (2002). Longitudinal impact of an inquiry-based science program on middle school students' attitudes toward science. *Science Education, 86*(5), 693–705.

Goodlad, J. I. (2004). *A place called school: Prospects for the future.* New York, NY: McGraw-Hill.

Guskey, T. R. (1989). Attitude and perceptual change in teachers. *International Journal of Educational Research, 13*(4), 439–453.

Guskey, T. R. (2000). *Evaluating professional development.* Thousand Oaks, CA: Corwin.

Guskey, T. R. (2002). Professional development and teacher change. *Teachers and Teaching, 8*(3), 381–391.

Harris, C. J., Penuel, W. R., D'Angelo, C. M., DeBarger, A. H., Gallagher, L. P., Kennedy, C. A., & Krajcik, J. S. (2015). Impact of project-based curriculum materials on student learning in science: Results of a randomized controlled trial. *Journal of Research in Science Teaching, 52*(10), 1362–1385.

Hewson, P. W. (2007). Teacher professional development in science. In S. K. Abell & N. G. Lederman (Eds.), *Handbook of research in science education* (pp. 1179–1203). Mahwah, NJ: Lawrence Erlbaum Associates.

Hewson, P. W., Kahle, J. B., Scantlebury, K., & Davies, D. (2001). Equitable science education in urban middle schools: Do reform efforts make a difference? *Journal of Research in Science Teaching, 38*(10), 1130–1144.

Horizon Research, Inc. (2015). University of Florida Unites Teachers to Reform Education in Science. Formative Evaluation Report. Chapel Hill, NC.

Johnson, S. M., Kraft, M. A., & Papay, J. P. (2012). How context matters in high-need schools: The effects of teachers' working conditions on their professional satisfaction and their students' achievement. *Teachers College Record, 114*(10), 1–39.

Knapp, M. S. (2003). Chapter 4: Professional development as a policy pathway. *Review of Research in Education, 27*(1), 109–157.

Knight-Bardsley, A., & McNeill, K. L. (2016). Teachers' pedagogical design capacity for scientific argumentation. *Science Education, 100*(4), 645–672.

Krajcik, J. S., Reiser, B. J., Sutherland, L. M., & Fortus, D. (2012). IQWST: Investigating and questioning our world through science and technology, (Middle School Science Curriculum Materials). Activate Science.

Ladson-Billings, G. (1994). *The dreamkeepers: Successful teachers of African-American children.* San Francisco, CA: Josey-Bass.

Lakshmanan, A., Heath, B. P., Perlmutter, A., & Elder, M. (2011). The impact of science content and professional learning communities on science teaching efficacy and standards-based instruction. *Journal of Research in Science Teaching, 48*(5), 534–551.

Lankford, H., Loeb, S., & Wyckoff, J. (2002). Teacher sorting and the plight of urban schools: A descriptive analysis. *Educational Evaluation and Policy Analysis, 24*(1), 37–62.

Lee, O., & Buxton, C. A. (2013). Teacher professional development to improve science and literacy achievement of English language learners. *Theory Into Practice, 52*(2), 110–117.

Lewis, E. B., Baker, D. R., & Helding, B. A. (2015). Science teaching reform through professional development: Teachers' use of a scientific classroom discourse community model. *Science Education, 99*(5), 896–931.

Lewis, S. C., Baker, N. D., & Jepson, J. C. (2000). *Critical trends in urban education: Fourth biennial survey of America's Great City Schools.* Council of the Great City Schools

Lotter, C., Rushton, G. T., & Singer, J. (2013). Teacher enactment patterns: How can we help move all teachers to reform-based inquiry practice through professional development? *Journal of Science Teacher Education, 24*(8), 1263–1291.

Loucks-Horsley, S., Stiles, K. E., Mundry, S., Love, N., & Hewson, P. W. (2010). *Designing professional development for teachers of science and mathematics.* Thousand Oaks, CA: Corwin Press.

Luft, J. A., & Hewson, P. W. (2014). Research on teacher professional development programs in science. *Handbook of Research on Science Education, 2*, 889–909.

Mesa, J., Pringle, R. M., & King, N. (2014). Surfacing students' prior knowledge in middle school science classrooms: Exception or the rule? *Middle Grades Research Journal: STEM Special Issue, 9*(3), 61–72.

Morrison, J. A., & Estes, J. C. (2007). Using scientists and real-world scenarios in professional development for middle school science teachers. *Journal of Science Teacher Education, 18*(2), 165–184.

National Research Council. (2012). *A framework for K-12 science education: Practices, crosscutting concepts, and core ideas.* Washington, DC: National Academies Press.

NGSS Lead States. (2013). *Next generation science standards: For states, by states.* Washington, DC: National Academies Press.

Paik, S., Zhang, M., Lundeberg, M. A., Eberhardt, J., Shin, T. S., & Zhang, T. (2011). Supporting science teachers in alignment with state curriculum standards through professional development: Teachers' preparedness, expectations and their fulfillment. *Journal of Science Education and Technology, 20*(4), 422–434.

Pelech, J., & Pieper, G. (2010). *The comprehensive handbook of constructivist teaching: From theory to practice.* Charlotte, NC: Information Age Publishing, Inc.

Penuel, W. R., Fishman, B. J., Yamaguchi, R., & Gallagher, L. P. (2007). What makes professional development effective? Strategies that foster curriculum implementation. *American Educational Research Journal, 44*(4), 921–958.

Pringle, R. M., Mesa, J., & Hayes, L. (2017). Professional development for middle school science teachers: Does an educative curriculum make a difference? *Journal of Science Teacher Education, 28*(1), 57–72.

Pringle, R. M., Mesa, J., & Hayes, L. (2018). Meeting the demands of science reforms: A comprehensive professional development for practicing middle school teachers. *Research in Science Education*, 1–29.

Ruby, A. (2006). Improving science achievement at high-poverty urban middle schools. *Science Education, 90*(6), 1005–1027.

Schneider, R. M., & Krajcik, J. (2002). Supporting science teacher learning: The role of educative curriculum materials. *Journal of Science Teacher Education, 13*(3), 221–245.

Schriver, M., & Czerniak, C. M. (1999). A comparison of middle and junior high science teachers' levels of efficacy and knowledge of developmentally appropriate curriculum and instruction. *Journal of Science Teacher Education, 10*(1), 21–42.

Schwarz, C. V., Reiser, B. J., Davis, E. A., Kenyon, L. O., Archer, A., Fortus, D., ... Krajcik, J. (2009). Developing a learning progression for scientific modeling: Making scientific modeling accessible and meaningful for learners. *Journal of Research in Science Teaching, 46*(6), 632–654.

Shepard, L. A. (2005). Linking formative assessment to scaffolding. *Educational Leadership, 63*(3), 66–70.

Tippins, D. J., Tobin, K. G., & Hook, K. (1993). Ethical decisions at the heart of teaching: Making sense from a constructivist perspective. *Journal of Moral Education, 22*(3), 221–240.

Tobin, K. G. (1993). *The practice of constructivism in science education*. Psychology Press.

van Driel, J. H., Meirink, J. A., van Veen, K., & Zwart, R. C. (2012). Current trends and missing links in studies on teacher professional development in science education: a review of design features and quality of research. *Studies in Science Education, 48*(2), 129–160.

Vescio, V., Ross, D., & Adams, A. (2008). A review of research on the impact of professional Learning communities on teaching practice and student learning. *Teaching and Teacher Education, 24*(1), 80–91.

Vygotsky, L. (1978). Interaction between learning and development. *Readings on the Development of Children, 23*(3), 34–41.

Whitworth, B. A., & Chiu, J. L. (2015). Professional development and teacher change: The missing leadership link. *Journal of Science Teacher Education, 26*(2), 121–137.

Wiggins, G., & McTighe, J. (2005). *Understanding by design* (2nd ed.). Alexandria, VA: Association for Supervision and Curriculum Development.

Wilson, S., Schweingruber, H., & Nielsen, N. (2015). *Science teachers' learning: Enhancing opportunities, creating supportive contexts*. Washington, DC: The National Academies Press.

Wilson, S. M. (2013). Professional development for science teachers. *Science, 340*(6130), 310–313.

Windschitl, M. (2002). Framing constructivism in practice as the negotiation of dilemmas: An analysis of the conceptual, pedagogical, cultural, and political challenges facing teachers. *Review of Educational Research, 72*(2), 131–175.

Yin, R. K. (2016). *Qualitative research from start to finish* (2nd ed.). New York, NY: The Guilford Press.

Yoon, K. S., Duncan, T., Lee, S. W.-Y., Scarloss, B., & Shapley, K. L. (2007). *Reviewing the evidence on how teacher professional development affects student achievement* (Issues & Answers Report, REL 2007—No. 033). Washington, DC: U.S. Department of Education, Institute of Education Sciences, National Center for Education Evaluation and Regional Assistance, Regional Educational Laboratory Southwest.

Zhang, X., McInerney, J., & Frechtling, J. (2010). Engaging STEM faculty in K-20 reforms – Implications for policies and practices. *Science Educator, 19*(1), 1–12.

Chapter 4
In the Mirror: Introducing Teachers to Practitioner Inquiry as Professional Development

As a science teacher educator, much of my research has occurred in the context of empowering teachers as continued learners. Most important, is the engagement of teachers in liberating practices that lead to awareness and improved knowledge, both necessary to identify and respond to problems of practice. In this chapter, I discuss how the teachers were introduced to and scaffolded to complete their practitioner inquiry as one component of the PD program. For all the teachers, this was their first intentional and systematic inquiry into their teaching. The chapter rests on the premise that even though teacher reflection is a natural part of teaching (Fischer, 2001; Loughran, 2002), and is grounded in teacher observation of their practice, PD should provide support settings that allow teachers to simultaneously study their teaching, their students, and themselves. In so doing, teachers construct local knowledge, question common assumptions, and build a conceptual framework to guide their teaching.

Educational research is a systematic and scholarly attempt to provide answers to the problems of teaching and learning within the formal educational framework (Cohen, Manion, & Morrison, 2008, p. 48). As a legitimate field of study, educational research provides insights into the procedures, rules, and principles relating to directing and transforming schools and classroom practices. However, in the realm and conduct of educational research, credence is rarely assigned to the role of teachers, who, as the change agents are responsible for shaping and transforming the culture of schools (Williams, Pringle, & Kilgore, 2019). Thus, while Cooley, Gage, and Scriven (1997) asserted that waves of educational reforms are mere trends in the literature, Scriven in the same AERA mini-feature article laments the failure of educational research to identify and improve on best practices (Cooley et al., 1997). This failure could be attributed to the positionality of classroom teachers in the process of traditional educational research. On the other hand, educators have argued for a form of educational research, grounded in the local concerns and complexities of the classrooms (Carr & Kemmis, 2009; Cochran-Smith & Lytle, 2009a, 2009b; Dana & Yendol-Hoppey, 2013) and which positions teachers as practitioner scholars.

R. M. Pringle, *Researching Practitioner Inquiry as Professional Development*,
https://doi.org/10.1007/978-3-030-59550-0_4

Teachers observe students, plan and enact opportunities to facilitate learning, and as learners, develop a repertoire of knowledge about teaching. Furthermore, effective teaching is informed by teachers' personal knowledge, trial and error, reflection on practice and conversations with peers (Fischer, 2001). In an era of reform in science teaching and learning, teachers are required to adopt and adapt new ways of practice. In such a realm, PD is seen as an avenue for challenging teachers' beliefs and supporting new learning about teaching and learning. PD providers have learned that supporting teachers' learning requires intentional connections to practice, ongoing engagement, explicit attention to disciplinary content knowledge in relation to contemporary beliefs about how students learn, and teachers engaging in studying their own practices (Battey & Franke, 2015; Kazemi & Franke, 2004). Each of these components of PD is essential in supporting teachers as continuous learners. Teachers develop new ways of thinking about teaching and learning science but more importantly, become better prepared to address essential issues of practice in their local context.

4.1 Introducing Practitioner Inquiry in Professional Development

At any level of the education system, we can safely declare that there is no shortage of research. However, in conversations with the teachers in this PD program, the general feeling was that research findings and best practices were mostly disconnected from the realities of their classroom practices. Ironically, this sentiment was also expressed by some educational researchers. Cooley et al. (1997) concluded that most traditional educational reforms occurred explicitly in the literature and on the pages of popular education publications and not in schools and classrooms. What is required therefore, is the compelling insider accounts of the complexities of teaching, learning, and schooling in today's globalized and test-based culture that emerge when teachers study their own practices (Cochran-Smith & Lytle, 2009a, 2009b).

Practitioner inquiry is largely about developing the professional dispositions of life-long learning and reflective and transformative practices (Stremmel, 2007; Stringer, 2014). The purpose of the practitioner inquiry as a component of the overall graduate degree program was twofold: to satisfy the university's program requirement as the capstone project in lieu of a thesis; and to develop the stance of inquiry among the teachers (Cochran-Smith & Lytle, 2009a, 2009b; Dana & Yendol-Hoppey, 2013). In this stance, and as life-long learners, the teachers' practices would transcend the usual patterns of thoughts to include a level of criticalness as they raise questions, and inquire into their practices. It is this inquiry stance that would support teachers' posture as educational researchers beyond the PD program. Just like "doing" science, practitioner inquiry is in response to questions and also about knowledge generation. It usually begins with an observation of a persistent issue or problem of practice that, as described by (Dana & Yendol-Hoppey, 2013), causes a sense of "wonder." Furthermore, seeking to redress emerging issues of practice, and as a continuous process, practitioner inquiry has the potential to ensure a cycle of continuous learning and informed teaching practices.

4.1.1 Practitioner Inquiry and Teacher Learning

Cochran-Smith and Lytle (2009a, 2009b), situating the process of teachers conducting inquiry under the conceptual umbrella of practitioner inquiry, forcibly uphold the importance of practitioners' knowledge, practice, and the change dynamics that occur in the range of professional contexts. They contend that "despite its variety, most versions of practitioner inquiry share a sense of the practitioner as knower and agent for educational and social changes" (p.37). The teachers, immersed in their sites of practice, are better positioned to confront their realities, transform their practices, and make the necessary adjustments for improved student learning. The emerging knowledge would be driven by local issues and concerns but the results are more likely to have immediate impact on practice. This is the first step toward knowledge and transformed actions in addressing the issues of classroom practices and students learning, and toward school change and reforms. Thus, it can be argued that transformed teaching practices are driven by teachers' desire to be more effective within the context of the complex nature of teaching and learning.

Teachers inquiring into their practices is not new and its importance in supporting their professional development has greatly evolved and gained prominence over the last 40 years. According to the Reflective Educator's Guide to Classroom Research (Dana & Yendol-Hoppey, 2009), practitioner inquiry is one of the three paradigms of educational research but is particularly focused on specific issues of practice as identified by the teacher. That is, practitioner inquiry is educational research. It emerges out of the teacher's concerns as opposed to being initiated by academia and consultants external to the school and classroom environments. Like all forms of research, teacher inquiry is a fundamentally social and constructive activity (Cochran-Smith & Lytle, 1993), but one which challenges the status quo of traditional research. When teachers engage in practitioner inquiry, the focus is on student learning as they examine their teaching practices within the context of their environment. In this position, teachers as learners, are in the cycle of constructing local knowledge, questioning common assumptions, and offering thoughtful critiques of the usefulness of educational research generated by others (Cochran-Smith & Lytle, 2009a, 2009b). Other colleagues, in separating practitioner inquiry from other educational research, highlight the conceptualization of the role of classroom teachers and the immediacy of the newly constructed knowledge in informing their practices. In the process, teachers as deliberative intellectuals, engage in systematic, intentional inquiry about their own practices and offer informed perspectives as they transform teaching and learning (Chandler-Olcott, 2002; Dana & Yendol-Hoppey, 2013). This systematic examination of practice and analysis of student's learning becomes interwoven into the daily and critical practices of teachers' reflection on their practices. The formalization of this reflection process according to Larrivee (2010), allows the teachers to move beyond a knowledge base of discrete skills to fit specific contexts, to a point where the skills are internalized enabling the emergence of necessary adjustments to practice.

4.2 Practitioner Inquiry as a Course

During the final two semesters of the degree program, through a blend of online and in-person one-day per month face-to-face sessions, the teachers were guided through the completion of a cycle of practitioner inquiry. The provision of the level of scaffolding within the context of a formal science education course was borne out of the need to ensure the timely completion of the program within the period of funding. Conducting a practitioner inquiry was a new experience for the group of teachers. In addition, accepting practitioner inquiry as educational research was at first a challenge partly because of its distinctiveness and reliance on the dual role of practitioner and researcher within the classroom. The course therefore allowed teachers to problematize their practice and identify appropriate questions to guide the process of inquiry. Teachers then engaged in the enterprise of data collection and analysis and the generation of knowledge, relevant and appropriate to their practices.

4.2.1 Introduction to Practitioner Inquiry

Classroom teachers are constantly faced with challenges in their teaching. Chief among the challenges are issues related to student learning. The teachers engaged in the PD program were being prepared to teach science in a manner consistent with the current reforms in science education. During the PD, teachers as continuous learners reflected on, practiced, and made sense of the new knowledge and experiences. Practitioner inquiry was introduced to the teachers as a systematic study of their professional practice with the goal of generating new knowledge pertinent to their context. At first, the teachers surfaced their understanding of the purpose of educational research and their perception of themselves as researchers. In small groups and later in whole group discussions, teachers responded to the following questions: In what ways can teacher as researcher enhance students' science achievement? And, what is the relationship between teaching and teacher research? In their responses to the questions, teachers recognized they occupied the best position to ask pertinent questions and make thoughtful adjustments to better facilitate students' learning.

4.2.2 Practitioner Inquiry as Educational Research

Course organization: The course was organized into eight modules over the duration of the final two semesters of the program. In each module, the teaching and learning activities included assigned readings, small and whole group forum discussions, journal reflection, and peer reviews. The first module was dedicated to unpacking the teachers' beliefs about the nature and role of educational research and served to introduce teacher inquiry as educational research. At first, the teachers resisted the conceptualization of practitioner inquiry as research. Their beliefs about research were directly associated with the use of rigorous scientific methods and as a process that "others" conduct. This was evident in their responses in their forum discussions in which they continuously raised questions about validity and reliability and their own capabilities as practitioners to generate knowledge. Course readings provided information that positioned teachers as agents for educational and social change (Cochran-Smith & Lytle, 2009a, 2009b), practitioner inquiry as a form of research designed to seek practical solutions to improve classroom practices (Stremmel, 2007), and a range of narratives from teachers who had conducted practitioner inquiries.

Six of the eight modules were dedicated to introducing the components of the process for conducting inquiry. The main text of the course, The Reflective Educator's Guide to Classroom Research: Learning to Teach and Teaching to Learn through Practitioner Inquiry by Dana and Yendol-Hoppey (2013), provided the framework for the timely introduction of the process. The text however was complemented by other readings. At each phase of the process of inquiry, course assignments required the teachers to write and revise drafts of their developing project. One of the early activities conducted in the course, required the teachers to surface their prior definition and beliefs about the role of traditional educational research. The teachers first responded individually and then in small groups shared and discussed their responses. While there were varied meanings of educational research among the group of teachers, a consensus was arrived at describing educational research as the process of generating knowledge in response to questions about teaching and learning and other areas related to education. The assigned readings and activities introduced practitioner inquiry as a disciplined process in which teachers carefully and systematically examined their own classroom practices using the techniques of traditional education research. These techniques included defining the issue or problem and the formulation of questions, collection of data related to the issue, and analysis, organization, and summary. Practitioner inquiry, however, does not end with the concise presentation of findings but leads to evidence-based practices and the generation of other questions. That is, the teacher is now engaged in the cycle of inquiry leading to improved practices in support of students' learning.

The course was taught primarily online and was supported by activities during the monthly cadre meetings. The cadre meetings were held mainly on the campus of the university and lasted over two days. During the cadre meetings, teachers were provided a space to share their experiences and collaborate as they work to make sense of the formal learning activities offered online. In addition to the cadre meetings and participation in small and whole group forum discussions, the teachers maintained an electronic journal that was shared only with the instructor. At the end of each module teachers completed application assignments related to the content. Some of the assignment tasks required the teachers to respond to questions that gauged their understanding or evoked reflection. Some examples of questions, what is the relationship between practitioner inquiry and professional development? In what ways is your classroom being impacted by your engagement in your practitioner inquiry?

The teachers' responses to the application assignments provided much insight into the development of their thinking over time. The responses also provided the topics for further engaged conversation with the instructor. At first the journal entries lacked depth as the teachers' responses appeared superficial and independent of their current experiences. Over time, the responses became substantive and teachers began to use specific examples to justify their ideas. For example, the following responses are representative of the evolution that occurred over time as the teachers conducted their inquiry.

Teacher 1:

> It can be difficult to be self-critical and one must be brave to face oneself in the mirror of inquiry. (Module 1)

> Everything came together for the inquiry project. The project was important; it was a big study. But just the idea that they taught us how to look critically at our own instruction, for us to ask, 'What can I do better?' that's not something you'd normally think of doing. (Module 7)

Teacher 2:

> I do not think there is much of a relationship between inquiry and professional development. (Module 1)

> Professional development is about learning how to be better at teaching science. My district holds professional development at the start of each school year … sometimes these meetings are not productive. Since being in the teacher inquiry course and examining the students who are not learning and what I need to change about my teaching, I see how teacher inquiry is professional development. I have learned so much about my teaching both from my students and from my peers going through the program with me. (Module 7)

At the end of the course, as teachers reflected on the process and its importance to teaching and to students' science learning, the following from a teacher's reflection is representative of the group's perspective. She wrote…

> A teacher inquiry project is more than merely reflecting on your practice, interactions with students, or daily lessons. I experienced teacher inquiry as a very systematic exploration of my own teaching and the effect of these teaching practices on my student. Interestingly for me, the process of teacher inquiry does not stop after the exploration of a specific wondering, it just raises more questions. So, it is really a continuous cycle that will continue to push

> me and my peers in this program to continue to examine our teaching, make changes, and be part of a body of research that influences reform in our classrooms (Elizabeth, Reflection submitted as a component of the capstone project).

Even though the practitioner inquiry was focused on each teacher's action within their immediate teaching context, there was a recognition of the importance of their findings to their immediate context and connections to the literature. In making a connection to traditional educational research via the literature, one teacher concluded, "Engaging in the inquiry process allowed me to connect on a deeper level to my teaching. I was also able to investigate if my wondering and practices were consistent with the literature. And it did – students' literacy skills are important in learning science."

4.2.3 From "Wondering" to Questions for Inquiry

The teachers were guided to take the first step in the process of practitioner inquiry during one of the early cadre meetings. They listed issues of their teaching and students' learning that had emerged during the enactment of the curriculum. Each teacher made a descending ranked-ordered list with the most pressing concern at the top. The teachers were then organized into teams of four based on the issue at the top of their list. In the group, they discussed the nature of the issue and identified the areas of teaching and learning that were being impacted. For example, teachers identified frustrations with low science achievement among populations of learners, the impact of low literacy proficiencies, and abilities to meet the learning needs of English Language Learners (ELL) in the inclusive science classrooms.

From the issues identified, teachers generated potential questions to guide their practitioner inquiry. The questions were then subjected to peer critique and evaluated in relation to the expressed "wondering," and guided by criteria identified by Ferrance (2000). The criteria were worded in the form of questions: Will the question require yes/no responses? Is there an immediate answer to the question? What kind of influence does the teacher have over the issue? Is the question relevant and appropriate and worthy of the time to be invested? To what extent will the response provide meaningful information? In small groups, teachers challenged each other and were encouraged to provide justification for each claim or question raised about the work of their peers.

4.2.4 How Do I Respond to My Questions?

The teachers having identified the question(s) and sub-questions, were now poised to make certain decisions before beginning data collection. The accompanied course readings provided guidance for developing their research plan and conducting the inquiry. The development of the research plan was an important aspect of the teacher inquiry process. The plan included a description of the area of focus – wondering, question(s) and sub-questions, rationale or purpose of the inquiry project; data collection strategies including teaching activities; and the timeline of research activities. In addition, upon conclusion of the research plan, the teachers were better able to articulate and justify the need to study their practices.

The teachers engaged in inquiry had to comply with the university's policy concerning research. They were introduced to the registration, review, and approval process of the institutional research review board (IRB). In the interim and before approval was granted by IRB, the course activity focused on exploration of the related literature and the nature of the data to be collected.

The process of literature review is recognized as central in advancing educational research. As researchers seek to build on and learn from prior research, the review of the existing literature allows for situating their interest within the broader scholarly and historical context and being able to distinguish what has been done to what needs to be done (Boote & Beile, 2005). While the practitioner inquiry was a response to the local classrooms, the teachers needed to develop familiarity with the knowledge base around their problem of practice. Teachers' exploration of the literature allowed them to learn from prior research on the topic and possibly added new insights – at least within their contexts. The teachers then wrote an annotated bibliography comprised of at least six sources from peer-reviewed journals of their choosing. This served as the first step in analyzing and synthesizing the information garnered from their field of interest. Situating their areas of interests in the existing literature served to provide a description and critical evaluation of the knowledge that was already existing in the literature. For example, one teacher concerned with the level of engagement and achievement of his English Language Learners (ELL), set out in the literature to better understand the relationship between science vocabulary and science learning among ELLs. His literature exploration led him to the use of cognates – the strategy he examined in his practitioner inquiry (Williams et al., 2019). In his reflection, the teacher shared that the literature provided productive insights into the relationship among ELLs' prior knowledge, their conceptual understanding, and the limitation of the focus on the science vocabulary.

After the IRB's approval was received, we entered into the planning phase which included the identification of the data and clarification of the process for collection and analysis. As teachers worked toward the development of the research plan, course activities guided them in the identification of the appropriate data to respond to their questions. They worked in assigned pairs to review and provide feedback to each other. Their responses to their peers were guided by the following prompts: To what extent could I seamlessly follow the plan as written to answer the questions posed? If I was to conduct this inquiry in my own classroom, what aspects of the plan as presented would I have issues following? What are some areas that lack clarity? Does the plan clearly outline how and what data will be collected? After the peer evaluation, the teachers revised their research plan guided by the feedback provided by their peers.

A review of the forum discussions periodically indicated that lingering struggles were occurring among some of the teachers. Teachers were sharing skepticism about their embrace of the practitioner inquiry and the nature of the data to be collected. In their forums, questions were raised such as: How do I set up the experiment? How will I change my notes into data? Some of these questions were directly related to aspects of traditional research. Even as they were immersed in the process, some teachers were still formulating their acceptance of practitioner inquiry as a viable educational research. This level of dissonance however became less overtime especially when teachers were engaged in collaborative analysis.

At this stage in the process, with the research plans in place, the teachers were ready to embark on data collection. However, just before they started data collection, Ashley, a recent graduate from of one of the educational doctoral programs in the college was invited to share her practitioner inquiry project with the group. In preparation for her visit to the cadre meeting, the teachers wrote the questions they were still having about the nature and credibility of practitioner inquiry as research. While many questions asked of Ashley were about the process, a number of teachers were interested in how her work were received among her peers and her administrators. In her presentation, Ashley reminded the teachers of the potential benefits of practitioner inquiry. She described her work with students who were placed in tiers 1–3, according to their reading levels. Students in tier three required the most support in their reading. She shared her data including a journal that was meticulously kept over the period. Other data sources included video clips of students' reading, transcripts of conversations among the students, and also her one-to -one conversations with individual students. Providing samples of her data sets, Ashley engaged the teachers in preliminary analysis. They read her data, identified hunches, and provided specific supporting data for each of their hunches.

In post cadre meeting reflections, teachers noted that while generalizability would be an issue for the "hard core" researcher, most of them shared they were now able to relate to the familiarity of what was described and explained in Ashley's research. According to one teacher's reflection, "Even though I am a science teacher, Ashley was describing things that have happened in my classroom with my students and I can see how that can be captured and shared."

4.2.5 In Collaboration: Learning with Peers

Embracing the stance of a teacher-researcher and participant observer, the teachers were now poised to gather fieldnotes, and explore the use of data collection tools described in their research plan. Some of the data sources included direct observation, journal entries, classroom artifacts such as students' classwork and notes, test scores, and memos. As the course entered the phase of data collection, the online platform was organized for instructor led tutorials and collaborative small group interactions. The platform allowed teachers to share their data and engage in the processes of analyses and interpretation. Each week, teachers made segments of their data available in the forum which were then read by the group members. Collaboratively they engaged in preliminary analysis and interpretation of each other's data. Furthermore, group members displayed a level of criticalness as they asked questions and offered suggestions. It was these suggestions, according to some teachers that upon reflection, were crucial in supporting them through the daily data collection and reflective processes. While the primary focus of the group interaction was on supporting the processes of analysis and interpretation, teachers' conversations also included clarifications of concepts related to teacher inquiry. For example, they discussed terms such as summaries and interpretation and differences between data and reflection and the selection of best evidence to support their claims.

Throughout the course, teachers shared their issues and concerns related to the process of data collection and ongoing analysis. The continued dialogue within the groups including the professor, helped to refine their thinking and brought to the fore sections of their investigated phenomenon that may have been overlooked. Teachers were also able to relate to the commonality among the experiences being discussed and that lent some credibility to their work. Once the teachers became immersed in data collection and analysis, their conversations within the groups began to show levels of ownership of their individual inquiry process. Some teachers began to ask questions of their practices and soon were noting that second guessing their actions and decisions was becoming a part of their daily teaching routine. According to Jennifer in her reflections "I can actually feel the shift in my thinking about my teaching…and the rapid number of questions that I am now asking of myself and my teaching." Susan a journalism major wrote, "who knew it would be such a mirror experience… I have actually developed a routine that allows me to see myself in real time making changes based on what I am observing and then writing the descriptions and my reflections and actions in my journal." The teachers were engaged in an intellectual pursuit in which, as the architect of their practice, made informed adjustments to their teaching toward supporting students' learning.

The process of practitioner inquiry includes aspects of traditional research. In practitioner inquiry, teachers investigate their practices and like traditional research, develop complex and refined understanding thus adding to the existing body of knowledge. For these teachers their practitioner inquiry was mostly an individual endeavor. Some scholars posit that teachers and the school systems are better served when practitioner inquiry is conducted collaboratively. The teachers in this program were situated at different geographical locations but bonded as learners in a PD program grounded in transforming their science practices. One aspect of the transformation was the development of a disposition toward inquiry that would allow them to raise questions about their practice and be vigilant in adjusting their teaching in response to their students' learning. Some would argue that raising questions is a common practice among teachers. However, in this course, teachers were prepared, as researchers, to engage in the cycle of inquiry seeking answers to questions, and generating knowledge to inform their local practices. Practitioner inquiry is a continuous process that involves periods of wondering and raising questions and a deliberate plan to reflect, observe, evaluate, and identify emerging questions. Engaging in the process therefore facilitate teachers as continuous learners where questioning their practice becomes a part of the daily deliberations and teaching culture (Dana & Yendol-Hoppey, 2013), and decisions about their practices are evidence-based.

The instructional strategies in the course were varied. The teachers interacted in small-and large-group discussions, shared observations of their classrooms, reflected systematically on their practices, and engaged in peer evaluation. The practitioner inquiry lead to much transformation in the teachers' practices. It facilitated the development of a level of sophistication in teachers' response to the issues they identified as "wondering" and which became the content explored in their first teacher inquiry.

4.3 Conclusion

The capstone projects submitted by the teachers contained the narrative of their lived experiences. The practitioner inquiry lead to the development of a stance that was reflected in their systematic reflection and transformed practices. It also facilitated the development of a level of sophistication in the teachers as they constructed the local knowledge in response to their issues of practice. While the project represented this particular teaching period, teachers were influenced by their past experiences which included multiple challenges in teaching and learning science. The practitioner inquiry provided the avenue to study and devise strategies to overcome the issues. At the start of the semester, each teacher identified a number of immediate issues arising from their professional practice and then made the executive decision as they selected the most pressing for their initial practitioner inquiry. Chapters 5, 6, and 7 present the voices of nine of the teachers in the PD program. The teachers, in real time, describe their experiences as practitioner researchers and the impact of the process on their learning as they surfaced and attended to local issues of practice and student science learning.

While these middle school teachers began the program with great concerns about their students' low achievement in science, they were aware of the long-term goal of their schools and districts. Their engagement in the PD would lead to a transformation of science teaching and learning in the partnering middle schools. During the PD program, the teachers were immersed in learning while enacting a curriculum that included recommendations for science learning and teaching in accordance with the current reforms in science education. The program activities deepened their science content knowledge and introduced the teachers to reform-based teaching practices. Yet, the teachers were still confronted with problems of practice in their classrooms. Their experiences in the PD program and their participation in the practitioner inquiry heightened their awareness about their role as researchers as they observe, frame, and study problems within their own practices and the impact on students' science learning.

References

Battey, D., & Franke, M. (2015). Integrating professional development on mathematics and equity: Countering deficit views of students of color. *Education and Urban Society, 47*(4), 433–462.

Boote, D. N., & Beile, P. (2005). Scholars before researchers: On the centrality of the dissertation literature review in research preparation. *Educational Researcher, 34*(6), 3–15.

Carr, W., & Kemmis, S. (2009). Educational action research: A critical approach. In S. E. Noffke & B. Somekh (Eds.), *Sage handbook of educational action research* (6th ed., pp. 74–84). London, UK: SAGE Publications Ltd..

Chandler-Olcott, K. (2002). Teacher research as a self-extending system for practitioners. *Teacher Education Quarterly, 29*(1), 23.

Cochran-Smith, M., & Lytle, S. L. (Eds.). (1993). *Inside/outside: Teacher research and knowledge*. New York, NY: Teachers College Press.

Cochran-Smith, M., & Lytle, S. L. (2009a). Teacher research as stance. In *The SAGE handbook of educational action research* (pp. 39–49). London, UK: SAGE.

Cochran-Smith, M., & Lytle, S. L. (2009b). *Inquiry as stance: Practitioner research for the next generation*. New York, NY: Teachers College Press.

Cohen, L., Manion, L., & Morrison, K. (2008). *Research methods in education*. New York, NY: Routledge.

Cooley, W. W., Gage, N. L., & Scriven, M. (1997). "The Vision Thing": Educational Research and AERA in the 21st Century Part 1: Competing Visions of What Educational Researchers Should Do. *Educational Researcher, 26*(4), 18–21.

Dana, N. F., & Yendol-Hoppey, D. (2009). *The reflective educator's guide to classroom research: Learning to teach and teaching to learn through practitioner inquiry*. Thousand Oaks, CA: Corwin Press.

Dana, N. F., & Yendol-Hoppey, D. (2013). *The reflective educator's guide to classroom research: Learning to teach and teaching to learn through practitioner inquiry* (3rd ed.). Thousand Oaks, CA: Corwin Press.

Ferrance, E. (2000). *Action research*. LAB, Northeast and island regional education laboratory at Brown University. *Educational Alliance, 34*(1), 1–33.

Fischer, J. C. (2001). Action research rationale and planning: Developing a framework for teacher inquiry. In G. E. Burnaford, J. Fischer, & D. Hobson (Eds.), *Teachers doing research: The power of action through inquiry* (pp. 29–48). Mahwah, NJ: Lawrence Erlbaum Associates.

Kazemi, E., & Franke, M. L. (2004). Teacher learning in mathematics: Using student work to promote collective inquiry. *Journal of Mathematics Teacher Education, 7*(3), 203–235.

Larrivee, B. (2010). What we know and don't know about teacher reflection. In E. G. Pultorak (Ed.), *The purposes, practices, and professionalism of teacher reflectivity: Insights for twenty-first-century teachers and students* (pp. 137–162). Lanham, MD: Rowman & Littlefield Education.

Loughran, J. J. (2002). Effective reflective practice: In search of meaning in learning about teaching. *Journal of Teacher Education, 53*(1), 33–43.

Stremmel, A. J. (2007). The value of teacher research: Nurturing professional and personal growth through inquiry. *Voices of Practitioners 2*(3), 1–9.

Stringer, E. T. (2014). *Action research* (4th ed.). Thousand Oaks, CA: Sage.

Williams, T., Pringle, R. M., & Kilgore, K. (2019). A practitioner's inquiry into vocabulary building strategies for native Spanish speaking ELLs in inquiry-based science. *Research in Science Education, 49*(4), 989–1000.

Part II
Introduction to Chapters 5–7: Researching Teachers Doing Inquiry: Presenting Their Stories

In the five-year federally funded professional development (PD) program, we prepared middle school science teachers to teach science in a manner consistent with reform in science education. One important aspect of the PD program was the support provided to teachers as they implemented the reform-based middle school science curriculum - Investigating and Questioning our World through Science and Technology (IQWST) (Krajcik, McNeill, & Reiser, 2008). Components of the PD program included: (a) the NSF-Math Science Partnership (MSP) Science Teacher Leadership Institute (STLI) - specially designed 2-year, science education graduate degree program, (b) workshops devoted to immersing teachers in the structure and design of IQWST as envisioned by the developers, (c) complementary PD activities that aligned with the formal courses as the teachers implemented IQWST and (d) district-level engagement and support. The two-year job-embedded STLI included graduate level courses specifically designed to deepen teachers disciplinary content knowledge and science specific teaching practices. In lieu of a thesis, the graduate program culminated with a capstone practitioner inquiry project.

When teachers conduct practitioner inquiry, they engage in a legitimate process of generating knowledge and insights about teaching and learning. The process is intentional and systematic and ultimately, the intent is to seek practical solutions to issues and concerns that emerge in their local contexts. Practitioner inquiry emerges from teachers' personal observations of their educational contexts which includes their effectiveness to facilitate students' learning and their ability to be critical of their practices. However, even though the classroom observations and emerging questions are context-specific, teachers as practitioner scholars, in studying their practices have the potential to link theory to practice (Bullough & Gitlin, 2001; Cochran-Smith & Lytle, 1993; Stremmel, 2007).

All nine practitioner inquiry projects presented in Chaps. 5, 6, and 7 represent a segment of persistent concerns in middle school science teaching and learning. The teachers were teaching a reform-based science curriculum. They were enacting science-specific teaching practices consistent with contemporary beliefs about how science learning occurs. Yet, they were experiencing high incidences of low performance in science achievement among populations of students. In response, immersed in the cycle of inquiry, the teachers illuminated the complex and challenging nature inherent in teaching as they moved from their traditional role as consumers to being generators of knowledge and understanding about science teaching and learning. Three topics of interest are presented in Chaps. 5, 6, and 7 – language and literacy, cultural relevance, and metacognition. In each chapter, three teachers allow us a view of their teaching and interactions with their students. Their project includes a description of their dilemma, the ways they examined their practice, and the adjustment made to their teaching to better address their problems of practice. Each chapter ends with a critical discussion of the meaning and importance of the teachers addressing the construct under study.

So, what have we learned from these practitioner inquiry projects?

Chapter 5: The ability to do and communicate science orally and in writing is important for science learning and for developing scientific literacy. The teachers understood the role and influence of language literacy on science learning in support of inquiry-based science. In Chap. 5, the voices of three teachers are elevated as they describe their dilemma and the ways they adjusted their teaching to better accommodate science learning of their students with low levels of reading and writing skills. The teachers present the impact of language proficiencies on science learning in their immediate contexts and the ways they examined their teaching thus making the necessary adjustments to increase their students' literacy skills.

Chapter 6: The subject of the teachers' narratives in Chap. 6 emerged in response to observed differences in academic achievement among diverse students. The teachers were grappling with the ongoing struggle for educational equity and access for all students identified as the most difficult educational and social challenge of the 21st century (NRC, 2012). To better serve the learning needs of the diverse students, teachers needed to value the diverse cultural experiences of their learners, develop their own cultural self-awareness, and understand the dynamics of cultural interactions that occur in the classrooms. It is this awareness that propelled the three teachers in this chapter to set out to examine their practices and develop their knowledge and teaching skills to improve their cultural competence.

Chapter 7: In chapter seven, the teachers embraced the importance of metacognition and its complement to inquiry-based science learning. They questioned the extent to which their teaching practices were effective and consciously explored strategies to facilitate the development of metacognitive skills among their students. The teachers approached the inclusion and development of metacognitive skills from three different strategic aspects. Collectively, they have shared images of their practices and necessary adjustments, the results of which can be adapted or at least will provide a framework for other middle school science teachers.

References

Bullough, R. V., & Gitlin, A. D. (2001). *Becoming a student of teaching: Linking knowledge production and practice* (2nd ed.). New York: Routledge Falmer.

Cochran-Smith, M., & Lytle, S. (1993). *Inside outside: Teacher research and knowledge*. New York: Teachers College Press.

Krajcik, J., McNeill, K. L., & Reiser, B. J. (2008). Learning-goals-driven design model: Developing curriculum materials that align with national standards and incorporate project-based pedagogy. *Science Education, 92*(1), 1–32.

Stremmel, A. J. (2007). The value of teacher research: Nurturing professional and personal growth through inquiry. *Voices of Practitioners, 2*(3), 1–9.

Chapter 5
Literacy Skills and Science Learning

5.1 Dilemma in Implementing a Reform-Based Science Curriculum When Students Struggle with Vocabulary and Reading Comprehension

Kim

I completed my undergraduate degree in science with the expectation of becoming a pharmacist. I detoured from pharmacy and became a certified middle school teacher through an alternative teacher education program. While I am comfortable with learning science content, I wrestle with how best to teach science to students for whom, among other issues, struggled with reading and writing. My students were persistently low academic achievers. I entered this professional development program with the hope of learning strategies and best practices that would increase the science achievement of my middle school students.

I teach science to sixth graders in a small rural county in the northeastern section of the state. Our middle school is situated in a run-down facility in one of the poorest sections of town. The school is consistently rated low in the yearly evaluation administered by the state and the students' science scores are persistently below the state's average. I entered the PD program hoping to be provided with tools and strategies to increase the science learning of my students from this side of town, who, historically are marginalized in science learning. I chose to conduct my teacher inquiry in my inclusive sixth-grade classroom. This school year, 11 of my 22 students in this class were, for the first time, mainstreamed into classes with regular students. The students' abilities range from very low to low-average as determined by their performances on district and state assessments. During their elementary years, they were taught in a self-contained classroom with others labeled as exceptional learners. Their classes had fewer students, but each student had varying disabilities. The children were also of different ages, and the academic focus was on reading and writing skills. This year, these students, placed in an inclusive classroom, are expected to make the necessary adjustments as they transition from

R. M. Pringle, *Researching Practitioner Inquiry as Professional Development*,
https://doi.org/10.1007/978-3-030-59550-0_5

elementary to middle school and, at the same time, learn science in a regular classroom with peers their own age.

I teach an inquiry-based science curriculum, developed around driving questions that play a central role in organizing students' learning. Driving questions are described by the curriculum developers as rich, open-ended questions that connect science learning with authentic interests and curiosities students have about the world. Inquiry-based science was different from the students' previous learning experiences. In fact, many of the students never had formal science lessons or learning science in inquiry-based lessons. In the first lessons of the school year, some students acted as spectators during the small group activities, while others attempted to engage with the materials. In a lesson on scattering and reflection, students used sensors to track light from a flashlight after it struck a mirror. I had nearly full class participation during this activity. Students became intrigued with their observation of light even though there were some issues with reading the instructions. As the lesson developed, the students were required to read a passage from their consumable text. The reading was intended to connect their observations from the class activity to real-world experiences. More importantly, a link among the in-class activities, reading, and homework tasks were designed to connect their previous knowledge to the new science ideas.

Once the reading task was introduced into the lesson, there was a dramatic change in the students' involvements. Many students became off task which led to issues of classroom management. My students were struggling with vocabulary and reading comprehension as well as expressing themselves in writing. I was concerned for their science learning. When I checked their reading scores from their school records, their scores were far below the average on state and district-wide reading assessments. While I was concerned, I knew the IQWST® (Investigating and Questioning our World through Science and Technology) curriculum I was teaching included many literacy components to support learners who struggled with the skills of reading and writing. In addition to being inquiry-oriented with emphasis on claims, evidence, and reasoning, the curriculum included the use of a Driving Question Board (DQB), suggestions to conduct meaningful discourse and integrating the use of the word wall in the development of each of the lessons. The DQB displayed the questions and sub-questions that guided the unit of study but also afforded students a space to display their questions as the lessons developed. Each lesson was also supported by reading materials to complement the science phenomena being studied. I felt that with the hands-on inquiry activities, embedded formative assessments, and the literacy strategies, I could help the struggling students excel in science *and* improve their reading, writing, and reading comprehension. Very soon, I realized that this was not going to be easy to achieve!

I observed my students enjoying themselves during the science lessons. They were observing and engaging in conversations and at times getting loud in their discussions. I enjoyed hearing the discussions that were occurring in the groups. However, I could not help but wonder about the struggles with the students, who, beyond the doing of the activities, were resisting involvement in the lessons. Furthermore, assignments that required reading were usually incomplete. As my concerns grew, I decided that for my teacher inquiry, I would explore how I could be more effective with the students whose science learning was being impacted by their low reading and writing skills. I wondered if focusing on the development of reading skills would increase their participation and learning in my science lessons. The more I thought about these students, I wondered if there were other strategies I could use to increase their vocabulary and reading comprehension. However, I did not want developing reading skills to replace learning science by observing and doing. How then could I teach reading in my science lesson and not lose the essence of good science teaching? I therefore set out to explore the literature to understand the existing research on developing the literacy skills of reading and writing among sixth graders in support of their science learning.

5.2 Exploring the Literature: Vocabulary

A review of the literature confirmed my thinking about the prominence of vocabulary in learning science. Armbruster, Lehr, and Osborn (2001) described vocabulary as words needed for effective communication. Students needed to develop a level of comfort with science vocabulary to be able to express their understanding of the concepts and ideas they were learning. This communication included being able to defend a scientific argument, write and read procedures, and express a meaningful understanding (Cromley, Snyder-Hogan, & Luciw-Dubas, 2010; National Science Education Standards, 1996). Many studies in the literature showed the connection between vocabulary knowledge and reading comprehension (Blachowicz, Fisher, & Ogle, 2006; Monroe & Orme, 2002; Nagy, 1988; Stahl & Fairbanks, 1986).

Vocabulary development is viewed as integrating experiences and providing sustained opportunities to struggling learners to use the words as they do and learn science. Although reading can increase a student's vocabulary, explicit instruction is required to improve contextual understanding (Armbruster et al., 2001; Bromley, 2007). Glowacki, Lanucha, and Pietrus (2001) found that the combination of direct and indirect vocabulary instruction led to significant improvement in the students' vocabulary knowledge and, in turn, reading comprehension. The link between vocabulary and comprehension is especially evident in science class where many words required for understanding the text, relevant literature, classroom activities, or teacher lecture are not encountered in everyday language. Additionally, science terms can have meanings that are different than the use of the word in everyday language (Rupley & Slough, 2010; Young, 2005), further adding to the issues. "The meanings are more restrictive and carry the concepts represented in the text" (Rubley & Slough, p. 100). In fact, words used in science class often embody the concepts themselves. If understanding of these terms is lacking, students will not only have trouble with conceptual understanding but may fail to be engaged in the process of learning science. Teaching vocabulary can create a link between the words used in science and the words used by the students. Additionally, vocabulary instruction can link science terms and concepts to background knowledge (Young, 2005) further supporting meaningful science learning.

Researchers have also established a strong relationship between vocabulary and student achievement (Cromley et al., 2010; Stahl & Fairbanks, 1986). They posit that deficiencies in vocabulary impact reading comprehension starting in the lower grades but with serious consequences for learning as the complexity of terms increase (Biemiller, 2003) usually as the students progress through the higher grades. The development of a rich vocabulary accompanied by conceptual understanding has the potential to lead to improved reading comprehension and academic achievement (Rupley & Slough, 2010; Taylor, Mraz, Nichols, Rickelman & Wood, 2009). In support of science learning, teaching strategies should seek to expand students' vocabulary. Educators caution however that such should be achieved within the context of engaged science learning as students connect their past experiences to the new and developing ideas during the examination of science phenomena (Greenwood, 2004). That is, learning experiences that develop science vocabulary within the context of inquiry-based science teaching has the potential to support students struggling with the skills of reading and writing. As I set out to examine and adjust my teaching, I selected and focused on strategies to support my students who were struggling with literacy. Guided by the literature, I selected text marking, graphic organizers, and cloze reading of passages. I also identified specific times in the development of the lessons that would be most appropriate to introduce the reading passages and the reading strategies.

5.3 The Cycle of Teaching Inquiry

The teacher inquiry project spanned three nine-week periods as I examined my teaching and tried to answer the following question: Will teaching reading strategies in inquiry-based science lessons improve the science learning of students struggling with reading comprehension and vocabulary? During the period of the project, the lesson topics included the nature of science, volcanoes, earthquakes, landforms, minerals, and rocks. I kept a journal and each day after school I described my teaching, observations of how the students were responding, and my reflections on the day's events. I also captured critical events that occurred as I worked closely with the 11 students. For example, I noted their responses during our interviews after each formative assessment and questions they may ask of their peers and as I interacted with them. In addition to my journal entries, the students' formative classroom-based assessments, unit tests and quizzes, and samples of their work provided the data for responding to my inquiry question.

5.3.1 *Data Analysis: A Process of Learning*

Each afternoon, I read and reflected on the journal entries of the day. After multiple reading, I used a color-coding scheme to identify the levels of students' engagement and their responses to the reading strategies. The coding scheme was usually followed by daily summary notes that captured what I was learning from the data. The notes focused on students' progress as they attempted to use or were successful in using the vocabulary related to the lesson. The learning from my summary notes were used to inform my practices in subsequent lessons. For example, I realized that the struggling readers were sources of distraction for their peers during small group reading activities. For the next lessons, even though I still spent more time with students struggling with reading comprehension and vocabulary development, I had to organize the student work groups in ways that allowed me more flexibility in monitoring and maintaining on-task behaviors.

To introduce the reading strategies, I had to first attend to the physical organization of the classroom in ways to accommodate the students working in groups. I made a deliberate decision to keep the teams intact as they worked during the inquiry-based activity and then transitioned into the reading instructions. As I observed and reflected on the progress in the small groups, the struggling readers were usually not engaged and were dependent on the other group members. I quickly realized, I had to exercise flexibility in my expectations within each of the group. The heterogenous grouping strategy based on reading proficiency was not supporting the struggling learners. In subsequent lessons, I deliberately re organized the teams to be more homogenous in their reading skills. The re configured groups allowed me much ease in differentiating my instructions and in introducing the reading strategies. The new grouping structure also allowed me more opportunities to deepen my relationships with individual students. As the relationship developed, students became comfortable as I interacted with them either individually or in their small groups. This allowed me to gain valuable insights into their learning needs as related to their reading skills. My observation of these interactions contributed to the decisions I made in adjusting my teaching.

5.3.2 *Vocabulary Instruction in Science Learning to Support Struggling Readers*

In the literature, it was clear that learning vocabulary words was important to learning science. For each lesson, I was mindful of the way vocabulary words were being introduced. Usually during the science activities, the students examined the phenomena, and words and terms would emerge during discussions. Having completed the activity, students, guided by their worksheet would continue in small group discussion as they write their claims and develop explanation. These activities were then followed by reading assignments. As the class on a whole continue the assigned classwork, I worked with the struggling readers introducing the readings that complemented the activities. During these interactions, I scaffolded the students to talk about what was done in the activity, allowed them to connect to other related experiences, and introduced the specific strategy that I selected. I deliberately did not focus on memorizing the list of words but expanded on the vocabulary as they emerged and within the context of the developing lesson. I chose four different mini-strategies to assist with enhancing my vocabulary instruction. The mini-strategies were surfacing and integrating students' prior knowledge, using a modified four-square graphic organizer, giving multiple exposure opportunities to the vocabulary terms, and incorporating activities that taught prefix and root words.

Surfacing and Integrating Students' Prior Ideas At the start of each new lesson, in general, I activated students' background knowledge by engaging in some small talk related to the topic and then posing questions. For example, I began a lesson on rocks by talking about the rock-climbing center in the mall. This was followed by questions that required them to share what they already knew about rocks. A few of them remembered that there were three types of rocks and that there was this cycle called the rock cycle. I elicited from them the definition of a cycle which we agreed on was a regular pattern of events that continued in the same order. Even with specific nudges, some students rarely responded during this section of the lesson. As the lesson developed, the students worked in cooperative groups to examine samples of rocks, described their observed differences, and participated in a simulation of the rock cycle. In one of the small group discussion one girl, one of the struggling learners, asked about the "purple looking stone" sometimes seen in science stores. This was definitely in reference to a field trip we had taken to the museum during our after-school program. After a teacher-led discussion, the students decided that because of the color, it was probably an amethyst which is a type of quartz. In my journal, I noted that these kinds of connections facilitated more meaningful discourse and usage of science vocabulary even among the struggling learners.

As I worked with the struggling readers in the small group, I also engaged their prior knowledge. Usually, I expected them to articulate what they may have learned or made sense of during the preceding science activity. In most cases, I had to use a lot of prompts – taking them step by step back through the recently conducted activity. In this way, my hope was that as they got into the text, they would be able to make connections between their class experiences and what was contained in the text. At first, it took more prodding to get a response from the students but over time, as they became more comfortable with my interactions in the small group, I had better responses from them even when they provided invalid answers. These times required more interactions and more scaffolding as they began to focus on the vocabulary and the comprehension of the text.

Using a Modified Four-Square Graphic Organizer This second mini-strategy proved to be most helpful in vocabulary instruction. As seen in Fig. 5.1, the science vocabulary was placed in the top left quadrant and the definition of the science in the top right square.

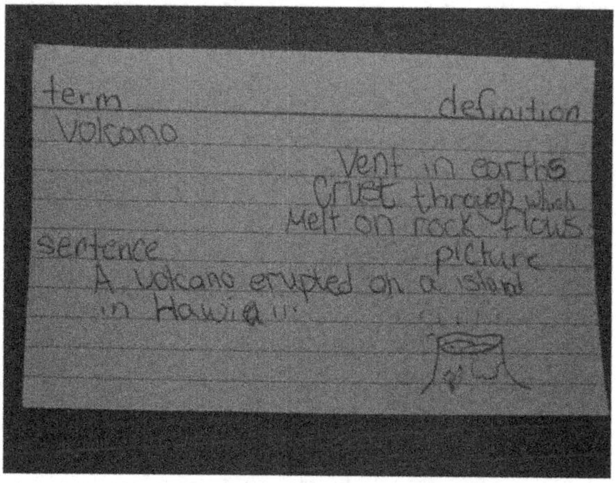

Fig. 5.1 Modified four-square graphic organizer

In the bottom left quadrant, students wrote a sentence using the science vocabulary term. In this quadrant, they were able to show their use of the vocabulary and how they made connections to the science content. In the bottom right quadrant, students used their creativity to represent their understanding of the concept. Most of them used drawings to represent their understanding. They enjoyed this strategy because they were allowed to express themselves in ways that were comfortable to them. The drawings also proved useful during class discussion. Students were encouraged to use their drawings as reference points during discussions.

Giving Multiple Exposure Opportunities to the Vocabulary Terms The third mini-strategy, the word wall, provided multiple opportunities to support students' use of the vocabulary words. The word wall consisted of a list of all the words and terms that were added as they emerged during the lesson or a set of lessons related to the learning objectives. I kept the word wall posted at the front of the room in an area where it was visible. Students had access and were encouraged to use the words as needed.

Incorporating Activities That Taught Prefix and Root Words My fourth mini-strategy incorporated the etymology of the vocabulary word. We broke down vocabulary words into prefixes, root words, and suffixes. For example, when we studied the five different systems of the Earth (atmosphere, geosphere, biosphere, hydrosphere, and cryosphere), we started by identifying the root word "sphere." We then discussed the possible definitions of the root word sphere, finally settling on "ball." For this activity, students made a foldable flipbook which included the word, a definition, and a picture. They created a comparative collection of all five systems of the Earth. The combinations of these four types of mini-strategies, as noted in my field notes, increased student engagement and their use of scientific terms in class discussions.

5.3.3 *Marking Text Paired with Cloze Reading Improved Reading Comprehension*

Being able to recognize vocabulary words was one task that had to be mastered toward improving student reading comprehension. To this end, later into the project, I introduced cloze reading strategies into the science lesson as I worked with the students identified as struggling due to their literacy skills. At first, I presented the students with a paragraph from the text in which some words were deleted. They were instructed to talk among themselves, discuss the science activity, and to fill in the missing words. This activity was overwhelming and caused some confusion for the students. I had to adjust my teaching. Each of the students needed assistance with the cloze reading strategy. After a few failed attempts, I revisited the cloze reading strategy. The students were assigned a sentence with a word deleted rather than an entire paragraph. The students were encouraged to talk among themselves and to fill in the missing word. The students still needed assistance with the cloze readings. I changed the strategy and introduced the students to marking the texts. In the small group, we read a section of text out-loud and as students re-read they were encouraged to highlight the bold-face and italicized words and definitions as well as underline main ideas of paragraphs. Eventually, I paired the cloze readings with the text marking as a teaching strategy for the entire class. These combined reading strategies helped to lower students' level of anxiety about reading, and I observed that with practice, students needed much less teacher assistance.

What Differences Did It Make? By the end of the first cycle of my teacher inquiry, students who were struggling with reading and using the vocabulary had begun to respond positively to the strategies introduced into the science lessons. Slowly the students began to show reduced anxiety related to their low reading and vocabulary skills. Even though they had not developed the reading skills on par with the grade level, the students no longer showed the kind of anxiety about reading that they did when they began their sixth-grade school year. I also noted changes during their performance in class. For example, my students had been dependent on me to supply the words when cloze reading was first introduced earlier in the project. As time progressed, I observed levels of independence and self-confidence in their approach to reading and efforts at connecting the words and searching the context of the readings. Furthermore, with scaffolding, there was marked improvement in their classroom assessments. My students also made positive comments such as "I want to stay in your class another block," "I enjoy doing this," and "these activities are fun." I saw a distinct transformation in the attitudes of these students toward text and its use in the science lesson to support their learning.

As Dana and Yendol-Hoppey (2009) discussed, teacher inquiry is a simple yet effective definition to a very complex, rewarding, transformative, provocative, and productive process. Teacher inquiry positions the teacher as the storyteller. The teacher becomes focused on providing insight into classroom practices in an effort to make a change. As I became immersed in the teacher inquiry, I could not help but stay laser focused on the impact of the strategies on the students. There were times when my observations became overwhelming and I questioned the extent of my effectiveness. Some students were very resistant to reading. At the beginning of my first teacher inquiry project, I had reservations about my ability to teach reading skills to middle school students during my science lessons. I recognized the impact of low reading skills on my students' engagement in my inquiry-based science lessons. I therefore explored reading strategies that would enhance students' capabilities as readers in science lessons. The inquiry allowed me to investigate, improve, and learn from my teaching practices, while seeking to support the science learning of students who struggled with literacy skills.

5.4 Conclusion

During my teacher inquiry, I examined and adjusted my science teaching in a manner that addressed my students who struggled with reading and writing. The students had a history of low achievement in the various subjects assessed at the district and state levels. In this first cycle of teacher inquiry, I chose to investigate the ways teaching vocabulary and reading skills can be integrated into science lessons. It was clear that the strategies I introduced offered levels of support to the struggling readers. It took time, however. It also required differentiated instructions to keep the entire class on task and focused on the lessons as they were being developed. The students showed varying responses to the reading strategies. They struggled with the cloze reading strategy. During the cloze reading, they developed a level of dependency on my assistance. In most cases, they constantly waited on me to provide the answers to any questions that were raised. I had to be deliberate in gradually reducing the level of scaffolding that I was affording them during this strategy. Nevertheless, as the struggling readers developed familiarity with the vocabulary words their reliance on my coaching became less over time. In addition, they started to show some responsiveness during the whole group activities and not only when they were examining science phenomena.

As I conducted the teacher inquiry, I became more focused in my reflection and intentional in adjusting my teaching to support my struggling readers in science learning. I realized that my science background had blinded me to the difficulties a student might have in reading, writing and talking about science. I was also handicapped because I am not a reading teacher. As I explored strategies to develop reading skills, I was guided by the research in the literature and support offered in the PD program. As a classroom teacher, I learned that strategies to increase reading and vocabulary skills can be integrated into science lessons. These learnings, although contextual and related to my classroom are essential as I continue to improve my practices. I have learned that even when integrating texts for reading in science, the experiences should be engaging, meaningful, and conceptually integrated into the immediate science activities. That is, the texts should make deliberate connections to science knowledge emerging from the related learning activities.

With this awareness I will continue to refine my use of these strategies and also continue the process of teacher inquiry. My hope is that my continuous participation in the process will make me better equipped to meet the needs of my struggling readers even as my wondering continues: Will the reading skills developed in my science lessons lead to better scores on external assessments?

5.5 Developing Students' Writing Skills: Toward a Deeper Understanding of Science

Ruth

I went to college to become an elementary teacher after driving a school bus for 10 years. As a bus driver in my community, I knew the students and their parents. If I had a problem on the bus, I could pull into a driveway and talk to the student's mom or granddad. After college, I got hired as a sixth-grade math and science teacher. One class in my bachelor's degree in education was about doing really great labs and having fun in science. I learned in that program how to get the kids interested and how to wow them with activities. As I started this program, my level of science knowledge was less than anyone else in my group. However, I have learned a lot about science and science teaching.

After driving the school bus for 10 years in a small, rural district in Northeast Florida, I returned to college to become an elementary teacher. After six years of attending college part time, I graduated with a Bachelor of Arts in Elementary Education. I now teach sixth-grade science and know most of the students from my years of being a driver in this rural school district. I did not plan to be a science teacher, but once I was given the assignment at the middle school, I developed an insatiable passion for learning science. Our county's eastern edge is the Atlantic Ocean, and while fishing is an occupation for a number of families, the major industries are tree farming, logging, and pulp mills. The county's population is 90% White, 6.14% Black or African American, and 3.8% Hispanic or Latino. The school has a population of 850 students and the demographics of the school is fairly consistent with the county's population. Approximately 55% of the students qualify for free or reduced-price lunch. In the recent annual state testing, less than 60% of students met minimum requirements for proficiency in reading, while approximately 50% met the minimum requirements for math.

The PD program was preparing and supporting our teaching of the IQWST curriculum which was being adopted by the school. The curriculum was developed in response to reform in science education. It included and required students to examine phenomena and make claims based on their experiences. During these inquiry-based science lessons, and as I walked around my classroom, my students were usually engaged in doing the science activity, which included observing the phenomena being studied. Sometimes the activities were completed individually, but most times the students worked in small groups. In this way, they supported each other as they shared their observations, generated claims, and used scientific reasoning to support their claims. After students completed the hands-on activities, I noticed they did not satisfactorily complete the related written assignment. These assignments, consistent with the curriculum, usually included writing a description of the procedure and evidence, claims, and explanation. When given opportunities to represent their learning in ways other than writing, the responses were usually more promising. For example, in an activity exploring smells in the air, students observed peppermint and were asked to draw a model of the odor. As my students learned about the particulate nature of air, they drew atoms and different arrangements of molecules showing carbon, hydrogen, and oxygen. They used letters to identify each molecule, along with a key to explain the parts of the model. The drawings also had arrows to show movement of the molecules. My students could show their understanding using two dimensional models but failed to adequately complete the accompanying writing assignment. During the lesson and observing my students defending their models, I was usually led to believe they had learned the material. When I read the written part of their assignments, their explanations did not reflect an understanding of the lesson. I sometimes wondered if I was grading the right papers! I thought about how I could address the disconnect between the students' understanding of science and their ability to write explanations or describe scientific phenomena.

5.5.1 My Wondering

I chose to do my teacher inquiry study with my fourth-period class made up of 80% girls and 20% boys. This was also my most racially diverse class with two African American girls and two Asian girls. My students, for the most part was engaged in the inquiry-based lessons that required them to make observations and share their ideas. However, they displayed poor responses to writing assignments related to the activities. When students write, a lot is revealed about their knowledge and ability to clarify their ideas. My students failed to complete their writing assignments associated with the science activities making it difficult for me to assess their learning. The activities in the curriculum used the claim, evidence, and reasoning (CER) framework which in most cases require writing. If I were only to consider the students' excitement when doing the activities, I would have concluded that they were learning. However, their incomplete written assignments and their refusal to represent their observations and claims was cause for concern. As I wondered about the impact of writing in science, I decided to focus my teacher inquiry on how I could best adjust my teaching to facilitate the development of writing skills among my students. The purpose of the teacher inquiry was, therefore to incorporate strategies that would develop writing skills among my sixth-grade students.

5.6 What Does the Literature Say About Writing to Learn Science?

One of the goals of science teaching, regardless of the science curriculum, is to teach students to become scientifically literate. When students are scientifically literate, they have a level of knowledge to make informed decisions and to participate in the discourse of the science community (Hand, Lawrence, & Yore, 1999; Norris & Phillips, 2003; Wallace, Hand, & Prain, 2004). This view of the nature of scientific literacy has several implications for the role and value of developing literacy skills while students are learning science. These literacy skills include reading, writing, listening, and talking. Writing promotes critical thinking skills and allows students to clarify their thinking (Baker et al., 2008) as well as actively construct new conceptual understandings (Wallace et al., 2004).

In a science lesson, students write reports, record experiments, and demonstrate their understanding of science concepts and practices (Glynn & Muth, 1994). Research has shown that when these writing practices are introduced during secondary science teaching, they become communication tools that will foster students' abilities to become scientifically literate. Writing as a literacy skill does not only support the processes of inquiry-based science. It represents the means through which students communicate their understanding to a wide range of audiences. Educators therefore encourage science learning to include students learning to write for different audiences including non-experts (Chinn & Hilgers, 2000). Rather than serving only as class assignments, teachers should also consider writing as a crucial problem-solving tool in the development of lifelong learners.

Teaching students' writing skills should be an integral part of the science lesson and not an additional task (Rivard & Straw, 2000; Wallace et al., 2004). Writing that is authentic and embedded into inquiry-based science activities requires less time to teach because students soon become motivated and involved in the process (Baker et al., 2008). I pondered over my concerns and the need to incorporate the development of writing skills into the inquiry-based science lessons. I realized that the first step in making the adjustment in my teaching would require me to modify the learning goals of the lessons to include the development of writing skills. This action heightened the need to include writing-to-learn activities into the instructional activities which reduced the constraints on the limited time given for each teaching period.

5.7 Where in the Inquiry-Based Science Lesson Do I Integrate Writing Skills?

While the literature offered a number of strategies to support the development of writing skills, I had to decide how best to integrate them into lessons that were already fully developed around specific learning goals. The learning goals of the curriculum were seamlessly developed in the instructional sequences, so I struggled with how to rewrite them to include writing skills as suggested in the literature. As I reflected on what I was about to try, it seemed as if I was about to teach students how to write and sacrifice other aspects of doing the science within the limited teaching period. I decided however that the best areas in the lesson to introduce the strategies to develop writing skills would be after examination of the phenomena, during the completion of the CER table, and the sense-making process. The students were already comfortable with doing the science activities and sharing their observations. In assessing their classwork and related assignments, the students had difficulties formulating and writing their claims and providing adequate written justifications. This was an important part in the inquiry-based lesson where students recognize evidence, make claims, and prepare to participate in the discourse of science.

5.7.1 Writing as Scientists: The Missing Link in a Sixth-Grade Class

The science lessons required students to work with science materials and, at times depending on the lesson, the Internet provided the phenomena for observation. Normally, as students entered the room, the learning goals were written on the white board and materials for the lesson were in the center of the tables. Each lesson began with the students surfacing their prior ideas either using the Know, Want, Learned (KWL) chart (Ogle, 1986), or responding to probing questions by writing their response on sticky notes. With the KWL, I would ask the students to identify what they already knew about the topic. For example, in a series of lessons on foods, we set out to explore the following question: Why do living things need food? In the first lesson, I had the students share reasons why animals needed food. As I conducted the class discussion, students raised other issues such as why different animals eat different things. I specifically wrote three of the responses on the white board under the title: What we already know. Over time and during the development of the lesson set, the students conducted a series of food tests toward identifying the nutrients in some familiar foods. Their student workbook had the table labeled claim, evidence, and reasoning (CER) which was also written on the white board. Since this was not the first time, the students used CER, they understood its role in writing their observation/evidence, claim, and reasoning. Students enjoyed the process of observing and to an extent enjoyed the discussions that required them to share their observations.

These sixth-grade students however, were not ready to fully participate in the process of writing to learn. This was important in their learning science but was fully expected during inquiry-based science. They struggled with writing their claim and even after small group discussions and probing, many students failed to complete the tables. I understood their issues with completing the reasoning component of the inquiry-based activity. Reasoning required a level of thinking and making connections as they moved from evidence-based claims. Even after the whole group discussions, many students were not able to complete the table on their own. As a result, I had to adjust my teaching using oral clues and, very slowly, worked with students to complete the writing aspect of the lesson. At times, I had to dictate or write on the white board to complete the CER table.

5.8 Moving Through the Process: From Doing to Writing

After the lessons on food tests and having established the presence of proteins, carbohydrates, and fats, we set out to answer the question – is water food? Students were instructed to make a list of five foods. During a whole group discussion, students shared one item from their list but were not to repeat what was already said. One student shared that water was a food and others agreed. The reasons given were that they had to drink a certain amount of water each day, and water was also listed on the food labels we had been using throughout the unit. After the discussion, I asked the class to revisit the lessons on the food test activities that were done over the last three lessons. Each table was to have one statement to respond to the question guiding the lesson: Is water food? The task required them to revisit their earlier lessons on food tests and observe labels from water bottles that had pure water. They observed that the water bottles had 0% nutrients and from the food tests water did not have any of the food substances. As a class, we agreed that water was not a food based on the evidence we had collected. Again, students had to be scaffolded as the discussion developed toward the completion of CER. Then, as we moved into discussing the reasoning component, students had difficulty making a statement that connected the claim to the wider scientific principle. The response was that food contained essential nutrients of proteins, carbohydrates, and fats; however, water was not food because it did not contain these nutrients. I realized that writing and expressing themselves after the science activity required more scaffolding. They needed the support to understand the activity beyond the fun and engaging in mixing and observing. My full dictation to complete the CER was not helping students to learn the science. They were just rewriting the correct responses. They were not thinking for themselves and were not making sense of the science. Some students were able to repeat some facts and were doing fairly well in the class assessment. Most of the students, however, were missing the evidence and were not able to make connections that would show conceptual understanding.

To keep the students focused on the science ideas as the lesson developed, I introduced the use of sticky notes to record key points that were then used to organize the students' thoughts and direct their writing. I thought that using sticky notes during class activities would help students when they had to engage in writing. Specifically, they had to write explanations in relation to the activity or responding to sense-making questions when making connections to the learning goals. Sticky notes became a regular feature in the science lessons and a way for students to capture ideas and key points that would be used in the writing assignments. Students also could write their thoughts or questions for later, or anything that made them think "WOW! I did not know that." They could abbreviate words, use texting language, and draw pictures. This was a form of free expression that would help them to clarify their thinking and develop their writing in a logical manner. After introducing the sticky notes and along with the sentence starters, I found more students were beginning, though slowly, to make some progress. They were starting to provide complete sentences and were citing evidence to support their responses. Upon reflection and as I reviewed the students' work I now had a better idea of the level of learning that was occurring.

The use of the sticky notes and the intense scaffolding to engage students writing began to impact the rubric in the curriculum that assessed students' learning. I thought that if I introduced sentence starters, I would be able to compare their learning against the rubric. The sentence starters were written on the white board to guide the students in completing their writing. I timed the sentence starters as the ideas were developed during the small or whole group discussions. However, students were not ready for that level of sophistication in their thinking. They were not able to translate into writing what was being learned in the inquiry-based lessons even with the sentence starters. They observed and talked about their observation but had to be constantly reminded how to generate their claims. I stressed certain points such as claims must be a complete sentence and should not begin with the words "yes" or "no," and evidence was the data collected through the senses or aided by scientific tools. Claims must be supported by evidence and, like scientists, we should avoid presenting feelings and opinions as evidence. One positive factor emerging from the whole group discussion and sharing of students' work was that the students constantly asked each other, "where is the evidence?" They were developing and using a key principle in inquiry-based science but were struggling with making connections to the science ideas.

My attempt at trying to improve students' writing by requiring them to express their learning changed how each lesson was developed. In the literature, when I learned about the different types of science writing, I was hoping to have the students write their observations, generate claims, and write the accompanying reasoning and explanation. By the end of the unit on foods, my expectation was that they would be writing for their parents or for their friends about the topic of food. I soon discovered that the students not only needed to learn how to write complete sentences, but inquiry-based science using the CER framework required a whole other set of skills. The students were excited to do the activities. They liked to share with each other and provide correct answers in class discussions. They, however, needed to understand the concepts and differences between claims, evidence, and reasoning and then be able to make the connection to the overall scientific principle. Upon reflection, I realized students needed further intense immersion with the strategies. Consistent with the literature, I concluded they needed multiple and consistent opportunities to better develop their writing skills (Applebee & Langer, 2009).

5.9 Final Thoughts and Future Questions

My teacher inquiry was about developing students' writing in ways that complemented science learning. The curriculum provided a clear pathway to developing the inquiry-based science lessons, but I soon learned that my sixth-grade students were not ready for some aspects of the curriculum such as writing to learn science. I introduced modeling cues, sticky notes, and sentence starters into the CER framework to provide scaffolding for the students to engage in writing. Introducing these strategies to the students required much scaffolding such as writing on the white board and dictating the information emerging from the lesson activities. Even just copying what was written on the white board was a challenge for some students. Some needed extended time, and others were not serious about the process of writing. I wanted all my students to be able to express themselves in writing to communicate their learning. Furthermore, the process of writing would help to clarify their ideas. I was forced to conclude that the issues with writing in science had to be tackled from different skill areas. There was the skill of expressing ideas using words and organizing ideas in ways that made sense in relation to the science content. Then students needed the skill of understanding and writing the vocabulary. Teaching students to do basic writing in inquiry-based science is a challenge requiring the teacher inquiry beyond a single quarter. Early in the process of the inquiry, I realized that to be successful meant I would need to systematically study my teaching every time I planned and taught a lesson.

By conducting this teacher inquiry, I discovered that the skill of writing was important in my own learning about myself as a teacher. As I did my writing, much of the learning about my teaching was revealed. My teacher inquiry resulted in changing my expectations and confronting those expectations in the contexts of what students can do. I learned the CER during the first course in the program and had additional experiences during the summer workshops as we learned about the curriculum. At first it was a struggle in my own learning – not the writing but communicating what was expected using my writing skills. Being able to formulate appropriate claims and provide the evidence and reasoning was a challenge for me during the PD, so I could relate to how the students felt. As a result of this inquiry-project, I feel empowered to question my teaching to better respond to the students' learning needs without sticking to the script of the planned instruction.

My teacher inquiry is ongoing. My submission for the master's program is complete but I will continue to be driven by my wondering to seek and to improve my practice. Upon reflection, I believe my students will benefit because I have learned about myself as a teacher and what I need to do to adapt my teaching so my students can be better science learners. It was a powerful feeling to see my students move from scribbling one or two words to providing complete sentences for explanation even if this occurred because of the intense scaffolding. These sentences gave me a window into their thinking as they formulated and expressed their science learning in words. As I move forward, my interactive use and combination of writing strategies will continue to inform my practice. I am hopeful, as researchers have suggested, that over time the students will be ready to write their thoughts and complete the activities as required without much prompting.

5.10 Literacy and Inquiry-Based Instruction: Accommodating Students with Special Needs

> Sean
>
> I was introduced to inquiry-based science teaching methods during coursework for my undergraduate degree. I really thought it was a great way to teach science because the students were using their senses and working to answer science questions. In the past year, I was assigned to teach science to two inclusion classrooms with special education students. The special education co-teachers, assigned to help with inclusion loved the idea of inquiry science. They thought the tactile and the experiential aspect of IQWST would meet the needs of the students.

In the recent state examinations, the students' reading scores at my middle school were below the state's average. The school's focus for this year was, therefore, on improving the reading scores among all populations of learners. I teach sixth-grade science in inclusive classrooms where some of my students are labeled as having learning disabilities (LD), are on the autism spectrum disorders (ASD), and other health impairments (OHI). Because of the diversity, I have been provided with a support teacher who assists both in planning and teaching. Each student was enrolled in exceptional student education (ESE) and had an individual education plan (IEP). IEPs are documents written to ensure that curriculum and classroom instruction are modified to best fit the needs of a particular student (Office of Special Education and Rehabilitative Services, 2000). In theory, the experiential learning in inquiry-based science lessons is well-suited for students with LD who often struggle with abstract concepts. However, students with LD need continual support such as coaching and scaffolding to help them make the necessary connections to the lessons. Students with ASD need support to work cooperatively with their peers. Students with OHI can have a wide variety of special needs requiring special accommodations. Over the years, I have struggled with students who were rarely completing their reading assignments even with assistance from the support teacher. Now I am teaching a curriculum that encouraged high student engagement. The lessons required students to create scientific models and work in groups to discuss and share their developing ideas. Despite these varied opportunities, many of my students with special needs struggled. Their levels of proficiency in literacy skills impacted their engagement in the broad array of science activities.

5.11 Wonderings

The ESE students had low reading skills and their academic achievements were below average when compared to their peers at the school. One of the issues I faced in teaching science was that the ESE students were not completing homework assignments that included reading. My observation revealed much frustration by the students especially when they received the grades assigned to their classwork. The difficulties ESE students were experiencing with reading comprehension impacted their learning in the inquiry-based-science lessons which required interactions with complementary text materials. My teacher inquiry project was initiated by my observation and wondering about the science learning of the ESE students – their reading scores, low achievement, and general deportment as science learners. For my teacher inquiry I selected the issue of supporting the development of reading comprehension skills among students classified as LD.

5.12 Related Literature

Educators have posited that experiential learning which is a hallmark of inquiry-based science is well-suited for students with LD. LD students struggle with abstract concepts and need continual support such as coaching and scaffolding to make sense of science ideas. Implementing inquiry-based science with students with LD is not new. Scruggs, Mastropieri, Bakken, and Brigham (1993) provided a history of the pedagogical wavering between a direct instruction and inquiry-based instruction towards science and special education. Inquiry-based learning has many benefits over direct instruction and its reliance on a textbook. It has been shown that students with LD struggle to read and comprehend information when compared to their grade-level peers without disabilities (Scruggs et al., 1993; Villanueva, Taylor, Therrien, & Hand, 2012). Brigham, Scruggs, and Mastropieri (2011) discussed that "students with [LD] have trouble acquiring information from lectures, class discussions, textbooks, and media presentations" (p. 224). In addition, the heavy reliance on textbook driven instruction and the rapid introduction of vocabulary would not be supportive of LD students' science learning (Brigham et al., 2011). With the discussed reading challenges experienced by students with LD, inquiry-based activities are advantageous over traditional education's textbook-centered focus. However, one concern about students with LD in inquiry-based teaching, is their processing speed compared to their peers. Mastropieri, Scruggs, and Butcher (1997) discussed that certain students with lower processing speeds can miss out on the important "Aha!" moment that can accompany inquiry-based science activities. They note, "the normally achieving students will have correctly 'constructed' the rule, but many (perhaps most) of the special education students will simple have been told the rule" (p. 208).

Inquiry-based science teaching can be effectively implemented with students from diverse cognitive backgrounds. According to researchers, its effectiveness is due to the high reliance on observing real world phenomena and collaboration with peers as they make sense of their experiences (Brigham et al., 2011; Easterly III & Myers, 2011; Jimenez, Browder, Spooner, & Dibiase, 2012; Scruggs et al., 1993; Villanueva et al., 2012). As I explored the literature, reading comprehension was one of the biggest challenges for students with LD (Brigham et al., 2011; Scruggs & Mastropieri, 2013), but when taught specific reading strategies such as annotating and completing text markings they were more likely to comprehend information from texts (Brigham et al., 2011; Kamalski, Sanders, & Lentz, 2008; Zywica & Gomez, 2008). In fact, Bakken, Mastropieri, and Scruggs (1997) found that teaching text structures to students with LD greatly improved their reading comprehension skills. This was the basis for my decision to introduce text marking strategies to my LD students during my inquiry-based science lessons.

5.13 My Cycle of Teacher Inquiry: Supporting Reading Among Students Labeled as LD

Learning science is one important content area for LD students. Because of the nature of inquiry-based science lessons, students learn about the world by participating in activities that allow them to develop scientific skills and practices. While the IQWST curriculum is not heavily reliant on text, students are still required to develop proficiency in reading. In the IQWST curriculum, texts are used to complement the science activities and provide connection to the wider science knowledge. Thus, it is important to ensure that students with LD develop skills in reading comprehension. I chose two strategies to use and study as I worked to develop the comprehension skills of my LD students. The strategies were text markings of reading passages and study guides to be completed after reading the text. These strategies are not included in the IQWST curriculum. I implemented these strategies with all 95 students in my sixth-grade science classrooms for a nine-week period but specifically focused on and systematically collected data from the 18 LD students.

For the teacher inquiry and over the nine-week period, I collected both quantitative and qualitative data. The data included student artifacts that included text markings, study guides, work samples from their workbooks, and assessments. I also included interviews with the support teacher, field notes taken in class, and IEPs of the students. Data analysis began as I examined my gradebook and recorded student grades from quizzes, comprehension checks, exams, and each text marking and study guide assignment. To determine the effectiveness of the strategies, I examined the students' workbooks and assessments. Informal assessments included comprehension checks, consisting of three or four questions based on the concepts and readings discussed in class during the previous few days. Formal assessments included IQWST quizzes and exams. I compared students' grades during the first and second quarter, before and after implementing the two new strategies. I also analyzed the average quiz, comprehension check, and exam grade for each student from both quarters. By analyzing the fourteen text marking assignments and three study guide assignments, I determined how often and with what efficacy the students were completing their new strategies. No easy answers emerged.

While this teacher inquiry partially relied on assessment data and grade averages, the focus was also on the ways I used the reading strategies during formative assessment. I read my reflections of the observations of the efforts being put into the work by the students. I also focused on the quality of the text marking strategies. Grades on second quarter assessments (Quizzes, Comprehension Checks, and Exams) were compared to first quarter assessments and referenced against three factors: Number of times the student demonstrated low effort on text markings, completion percentage of text marking assignments, and completion of study guides. I then examined the level of completion for study guides and text markings as compared to the performance on assessments.

I introduced text markings by demonstrating the process as I worked through one passage of text. For markings to be useful, a student was expected to mark sections of the text with symbols to denote an idea or concern. For example, in the margin next to a section of text, a student placed a "**?**", to ask a question; "**!!!**", for an interesting fact; "**E**" for Evidence; "**>**" a check mark for *I knew that* (prior knowledge); "**N**" for new information; **<>** Reason looks important; and Box It for vocabulary. I modeled the strategy, then read another section of text out loud to the group; we marked the text together. After modeling and group work, I asked students to independently mark the text for the lesson. Students were taught this process for two reading assignments during class. I also checked their reading homework assignments, monitoring for the correct use of the text markings on student workbooks. They completed a total of 25 reading assignments. For data purposes, I assigned ratings noting the extent to which students had completed the text markings with correct usage throughout the reading.

5.13.1 Text Marking Efforts Varied: Increased Effort Meant Increased Engagement

Introducing the text markings allowed me to better identify students' progress as they completed the reading and comprehension required during their homework assignments. Completion of the text markings also gave me a window into the students' understanding of the science ideas being taught. Text marking strategies became important, regardless of the reading comprehension level of the student using them, to demonstrate the students' engagement with the content in their reading. Students who conducted the text markings improved their science comprehension. In addition, the text markings became an extremely useful strategy to monitor students' effort.

5.13.2 Text Markings Improved Reading Comprehension

I found a notable difference in assessment performance of students who completed 50% or more of their text marking assignments to those who completed less than 50%. Students who consistently completed their text marking assignments demonstrated greater comprehension of science concepts. My near-daily progress monitoring of my students over the course of the second quarter provided enough data so that I could say, with confidence, that the text marking strategy was effective in improving the performance of ESE students in my classroom.

5.13.3 Students Adapted Text Marking Strategies in Personal Ways

Students embraced the text marking outline I provided on their bookmarks and adapted it for their use. They used highlighters and underlined information to further their use of this reading comprehension strategy. Some students found that highlighting and underlining information were more useful for them than making the prescribed text marking. Other students focused on boxing important information and vocabulary which was not required. This was an unexpected result and was noted throughout my assessments and my analysis of their workbooks. The students' responses were evidence that teaching them the strategy to analyze a reading passage was a worthy endeavor in supporting their learning. Students taking a technique and retooling it to best fit their own learning skills were great finds during my teacher inquiry.

5.13.4 Study Guides Did Not Improve Science Vocabulary and Decreased Some Students Scores

As the lessons developed I introduced study guides to support the learning of science vocabulary. I was careful to ensure that the vocabulary words emerged during the activities and also during the discussion. For the LD students, I only required them to associate the words with the concept within the context of the lesson. The vocabulary study guides did not achieve the level of success I was anticipating. For example, students who completed 90% or 100% of the three study guides showed that they were putting forth a good amount of effort to learn the vocabulary. Despite their efforts, I could not find evidence that completing study guides improved their science vocabulary or reading comprehension. I realized that many students Googled the answers or searched the Internet instead of reading the text. In fact, high completion rates of the study guides, in the 90–100% range, seemed to negatively impact some outcome measures. The extra time spent on study guides may have led to less time working on other homework assignments, lowering learning outcomes. As the year progressed, I stopped using study guides. Further observation of the students' progress and the effectiveness of the study guides lead me to abandon this strategy for use with the LD students.

5.14 Discussion

My wondering was about the extent to which I could improve the reading skills of my LD students during inquiry-based science teaching. My observation revealed that reading comprehension was a weakness in the science learning of the students with LD (Brigham et al., 2011; Scruggs & Mastropieri, 2013) even though I was teaching inquiry-based science. I researched the literature for strategies to incorporate into my teaching. I therefore implemented text markings and had some success with my LD students as they identified the main topic and important information in a passage. Zywica and Gomez (2008) showed that annotations and text markings were useful reading strategies, and the findings of my teacher inquiry supported their claim.

I also found that vocabulary study guides had a small effect on student learning and was ineffective in promoting meaningful understanding during vocabulary acquisition. On average, the focus students completed their study guides at a rate of under 50% which was reflected in their homework grades and overall performance in class. Moving forward, I will need to look at other options to support the science learning of my LD students with a focus on developing their literacy skills related to vocabulary. Inquiry-based education focuses on acquisition of knowledge through active engagement with real world phenomena and not memorization of vocabulary words (Brigham et al., 2011). The vocabulary study guides supported memorization and was not supportive of the LD students.

Each student with LD is unique. Classification of disability, even within groups of similar disabilities, cannot be blanket statements applied to all but guideposts to assist (Office of Specical Education and Rehabilitative Services, 2000). Where one student with a certain type of LD might need assistance or enrichment in a particular area of inquiry-based science education, other students with the same LD might not. My students with LD were low performers. They required more intervention than was provided by the reading text markings and vocabulary study guides during the time allotted for the teacher inquiry. My personal goal for the teacher inquiry was that the LD students would greatly benefit from the adjustments I made to my teaching. As I analyzed the data, it became clear to me that more interventions would be required to help LD students improve their reading and science comprehension to an appreciable level. However, teaching text marking strategies was important – responses of the students varied at different times during the course of instruction.

5.15 The Promise

As Dana and Yendol-Hoppey (2009) discussed, taking my own path towards professional development with comprehension strategies provided a sense of ownership of the occurrences in my science teaching. The students classified as LD struggled with the idea of reading assignments that complemented inquiry-based science learning. Inquiry-based science teaching focuses on the development of knowledge through the learners' experiences. The experiences include asking questions, examining science phenomena, and generating explanations by developing evidence driven claims. My students needed support in developing their vocabulary skills. I found that some traditional study guides did not support my LD students' learning of vocabulary even when used in the context of the developing science lesson. Text markings, however, made sense to the students. As my LD students struggle with reading comprehension and the meaningful development of science vocabulary to communicate their learning, I learned that my cycle of teacher inquiry will need to be a staple in my teaching and interacting with my students. The teacher inquiry allowed me to address my own capabilities. I was able to make the necessary changes to my teaching to meet the learning needs observed in my classroom. While I did not realize the full extent with students' learning, conducting the teacher inquiry was an important turning point in my professional growth. The teacher inquiry processes of observation, documentation, reflection, and decision to act and make changes provided a new level of comfort as a teacher – as I use evidence-driven claims and reasoning to address my problem of practices.

5.16 Researching Practitioner Inquiry: Literacy Practices and Science Learning

The international spotlight on science literacy has heightened the approach and the visibility of the place of literacy skills within the context of K-12 science education. The discipline-specific literacy practices of reading, writing, speaking, and listening are integral components of curriculum in many developing and developed countries. For example, outside the U.S., the skill of reading comprehension is considered a measured component of the performance of International Student Assessment (Alvermann & Mallozzi, 2009; Patterson, Roman, Friend, Osborne, & Donovan, 2018). In the U.S., the literacy skills are critical to building knowledge in science (NGSS Lead States, 2013), as teachers introduce and scaffold students into the discipline-specific ways of producing science knowledge. For years, and even amidst reform efforts, science teaching has persistently failed to effectively engage learners in ways consistent with how scientists do science. The textbook-driven approach to teaching misses the unique mix of inquiry and argumentation (Yore, Hand, & Florence, 2004), resulting in science learning as an accumulation of facts. When science is presented as finished products, independent of the processes and human aspects of knowledge construction, learners develop an incomplete understanding of the nature of science. As a result, science achievement becomes measured by the extent to which students are able to recall snippets of science information.

Literacy skills – listening, talking, reading, and writing either individually or in combinations are important for science learning and for developing scientific literacy. It is safe to declare that this global visibility is in part supported by an understanding of the place of literacy skills in the work of scientists. Literacy skills are foundational to the work of scientists. Scientists examine phenomena, propose explanations based on evidence, construct theoretical explanations, engage in scientific argumentation, and communicate scientific knowledge, principles, and procedures (Howes, Lim, & Campos, 2009; Norris & Phillips, 2003; Wallace et al., 2004; Yore et al., 2004). Scientists use these literacy skills before, during, and after their process of inquiry. The ability to do and communicate science orally and in writing is therefore important for both the work of scientists and for science learning. This is supported by an established body of evidence from educational research that have examined the centrality of literacy skills in doing and learning science. Researchers have therefore proposed strategies and emphasized the importance for science teachers to have a professional base of knowledge to deal with the challenges of literacy in science learning. Teachers have been the recipients and consumers of the knowledge garnered from these researches. However, missing from the literature is the perspectives and first-hand accounts of teachers' lived experiences with the challenges of literacy skills in middle school science classrooms. The three narratives in this chapter provide insights into classrooms as science teachers confront low proficiencies in literacy skills among their struggling learners.

The severity of low proficiency in literacy skills among struggling learners though not unique is a critical challenge for middle school teachers (Patterson et al., 2018; Yore et al., 2004). The teachers' narratives revealed that they intuitively recognized that effective reading and writing skills were critical to their students' achievement. They were teaching a middle school curriculum that privileged inquiry-based science teaching and which relied on students experiencing science in a manner consistent with how scientists conduct their practices. Their inquiry-based science lessons included the examination of phenomena, and the utilization of evidence to address real world questions. The curriculum provided students the opportunity to actively participate in practices as they learn science – a stark difference from the more traditional lecture and textbook-driven approaches to science teaching. In addition, learning was further enhanced as students recorded and shared their ideas, generated evidence supported claims, and engaged in scientific discourses. These literacy practices, as tools for content area learning, also allow for the added benefit of the socialization of students into the discourse of science (Pearson, Moje, & Greenleaf, 2010). Scholars contend that when students have difficulties with literacy skills they are usually limited in discipline-specific learning (August, Branum-Martin, Cardenas-Hagan, & Francis, 2009). Clearly, middle school content area teachers are faced with a significant challenge, given the disconcerting low levels of language literacy skills among their students.

The middle grade science teachers in this program had a legitimate concern about developing the literacy skills of their students. Consistent with the literature, they understood the role and influence of literacy skills on science learning. The deficiency in literacy skills hampered students' science achievement and placed them at risk of educational failure in science (Francis, Rivera, Lesaux, Kieffer, & Rivera, 2006), and other subject areas. At the beginning of the school year and confirmed by the scores on the state's assessments from the previous year, teachers recognized that many of their students struggled with reading and writing. As the teachers enacted the curriculum, they observed that students struggling with language and literacy skills were examining science phenomena and were developing and using science and engineering skills. There was an appreciable level of engagement as the students participated in the science activities. A pattern of findings, consistent with research, indicated a decrease in off-task behaviors as involvement in science activities increased. This was a notable observation made of all students including those whose language literacy skills were below grade level. The teachers were however frustrated, when students' achievements as assessed on formative and summative assessments were low due to their levels of literacy skills. These students were not able to participate in aspects of the lessons that required the practices of thinking about, talking about, reading about, and writing about their science experiences.

In making the adjustments to their pedagogical practices, the teachers organized their lessons to include the repertoire of strategies suggested in the literature. The strategies were carefully integrated into the lessons and were not merely added to the instructional sequence. At the most basic level, they set out to teach the literacy skills in ways to help learners make connections to what they already knew or had experienced during the inquiry learning activities. Notable, the texts associated with the lessons were specifically developed to complement the science ideas. The middle school teachers were well-versed in their content knowledge and inquiry-based teaching. They however experienced much difficulty delivering the quality literacy instruction to effectively develop the skills in the context of their science curriculum. In addition, the teachers recognized that learners who were deficient in literacy skills required instructions based on their respective level of proficiencies. To fully immerse struggling learners in meaningful science learning experiences in an inclusive classroom, requires differentiated strategies to accelerate both the development of literacy skills and facilitate the conceptual understanding of the scientific knowledge.

The teachers as content area experts acknowledged that incorporating the specific development of the literacy skills during science lessons was about using the skills as tools for thinking and learning. According to researchers, teaching in the content area such as science should not be just about basic literacy skills. Deep learning in science requires students to enact new identities as they read, write, speak, and make meanings in different ways (Faller, 2018; Yore et al., 2004). As teachers conducted the teacher inquiry projects, they reflected and were intentional in adapting research-based writing and reading strategies to support their students' science learning. Their narratives revealed the status of literacy skills among their struggling learners and the challenges and impact on science teaching and learning. The teachers made some progress but evident in their experiences was the need for guidance and support in the development of the requisite knowledge needed to teach such skills within the context of their science lessons. While the teachers provided a window into their classrooms and the impact of practitioner inquiry on their practices, their findings have implications for teacher education. The curriculum in teacher education programs, should include a knowledge base that allows teachers to diagnose and enact effective instructional strategies to address literacy skills within the context of science learning.

References

Alvermann, D. E., & Mallozzi, C. A. (2009). Moving beyond the gold standard: Epistemological and ontological considerations of research in science literacy. In M. Shelley II, L. Yore, & B. Hand (Eds.), *Quality research in literacy and science education* (pp. 63–81). Dordrecht, The Netherlands: Springer.

Applebee, A. N., & Langer, J. A. (2009). EJ extra: What is happening in the teaching of writing? *The English Journal, 98*(5), 18–28.

Armbruster, B. B., Lehr, F., & Osborn, J. (2001). *Put reading first: The research building blocks for teaching children to read.* Washington, DC: National Institute for Literacy.

August, D., Branum-Martin, L., Cardenas-Hagan, E., & Francis, D. J. (2009). The impact of an instructional intervention on the science and language learning of middle grade English language learners. *Journal of Research on Educational Effectiveness, 2*(4), 345–376.

Baker, W. P., Barstack, R., Clark, D., Hull, E., Goodman, B., Kook, J., … & Weaver, D. (2008). Writing-to-learn in the inquiry-science classroom: Effective strategies from middle school science and writing teachers. *The Clearing House: A Journal of Educational Strategies, Issues and Ideas, 81*(3), 105–108.

Bakken, J. P., Mastropieri, M. A., & Scruggs, T. E. (1997). Reading comprehension of expository science material and students with learning disabilities: A comparison of strategies. *The Journal of Special Education, 31*(3), 300–324.

Biemiller, A. (2003). Vocabulary: Needed if more children are to read well. *Reading Psychology, 24*(3–4), 323–335.

Blachowicz, P., Fisher, D., & Ogle, S. (2006). Vocabulary: Questions from the classroom. *Reading Research Quarterly, 41*(4), 524–539.

Brigham, F. J., Scruggs, T. E., & Mastropieri, M. A. (2011). Science education and students with learning disabilities. *Learning Disabilities Research & Practice, 26*(4), 223–232.

Bromley, K. (2007). Nine things every teacher should know about words and vocabulary instruction. *Journal of Adolescent & Adult Literacy, 50*(7), 528–537.

Chinn, P. W., & Hilgers, T. L. (2000). From corrector to collaborator: The range of instructor roles in writing-based natural and applied science classes. *Journal of Research in Science Teaching: The Official Journal of the National Association for Research in Science Teaching, 37*(1), 3–25.

Cromley, J. G., Snyder-Hogan, L. E., & Luciw-Dubas, U. A. (2010). Reading comprehension of scientific text: A domain-specific test of the direct and inferential mediation model of reading comprehension. *Journal of Educational Psychology, 102*(3), 687.

Dana, N. F., & Yendol-Hoppey, D. (2009). *Facilitator's guide: The reflective educator's guide to classroom research: Learning to teach and teaching to learn through practitioner inquiry.* Thousand Oaks, CA: Corwin Press.

Easterly, R. G., III, & Myers, B. E. (2011). Inquiry-based instruction for students with special needs in school based agricultural education. *Journal of Agricultural Education, 52*(2), 36–46.

Faller, S. E. (2018). Reading and writing as scientists? Text genres and literacy practices in girls' middle-grade science. *Journal of Adolescent & Adult Literacy, 61*(4), 381–390.

Francis, D. J., Rivera, M., Lesaux, N., Kieffer, M., & Rivera, H. (2006). *Practical guidelines for the education of English language learners: Research-based recommendations for instruction and academic interventions*. Portsmouth, NH: RMC Research Corporation, Center on Instruction.

Glowacki, D., Lanucha, C., & Pietrus, D. (2001). *Improving vocabulary acquisition through direct and indirect teaching*. Washington, DC: Distributed by ERIC Clearinghouse. https://eric.ed.gov/?id=ED453542

Glynn, S. M., & Muth, K. D. (1994). Reading and writing to learn science: Achieving scientific literacy. *Journal of Research in Science Teaching, 31*(9), 1057–1073.

Greenwood, S. (2004). Content matters: Building vocabulary and conceptual understanding in the subject areas. *Middle School Journal, 35*(3), 27–34.

Hand, B., Lawrence, C., & Yore, L. D. (1999). A writing in science framework designed to enhance science literacy. *International Journal of Science Education, 21*(10), 1021–1035.

Howes, E. V., Lim, M., & Campos, J. (2009). Journeys into inquiry-based elementary science: Literacy practices, questioning, and empirical study. *Science Education, 93*(2), 189–217.

Jimenez, B. A., Browder, D. M., Spooner, F., & Dibiase, W. (2012). Inclusive inquiry science using peer-mediated embedded instruction for students with moderate intellectual disability. *Exceptional Children, 78*(3), 301–317.

Kamalski, J., Sanders, T., & Lentz, L. (2008). Coherence marking, prior knowledge, and comprehension of informative and persuasive texts: Sorting things out. *Discourse Processes, 45*(4–5), 323–345.

Mastropieri, M. A., Scruggs, T. E., & Butcher, K. (1997). How effective is inquiry learning for students with mild disabilities? *The Journal of Special Education, 31*(2), 199–211.

Monroe, E. E., & Orme, M. P. (2002). Developing missing vocabulary. *Preventing School Failure, 46*(3), 139–142.

Nagy, W. E. (1988). *Teaching vocabulary to improve reading comprehension*. Urbana, IL: National Council of Teachers of English.

National Academy of Sciences & National Research Council. (1996). *National science education standards*. Washington, DC: National Academy Press.

NGSS Lead States. (2013). *Next generation science standards: For states, by states*. Washington, DC: National Academies Press.

Norris, S. P., & Phillips, L. M. (2003). How literacy in its fundamental sense is central to scientific literacy. *Science Education, 87*(2), 224–240.

Office of Specical Education and Rehabilitative Services. (2000). *A guide to the individualized education program*. Jessup, MD: U.S. Department of Education.

Ogle, D. M. (1986). KWL: A teaching model that develops active reading of expository text. *The Reading Teacher, 39*(6), 564–570.

Patterson, A., Roman, D., Friend, M., Osborne, J., & Donovan, B. (2018). Reading for meaning: The foundational knowledge every teacher of science should have. *International Journal of Science Education, 40*(3), 291–307.

Pearson, P. D., Moje, E., & Greenleaf, C. (2010). Literacy and science: Each in the service of the other. *Science, 328*(5977), 459–463.

Rivard, L. P., & Straw, S. B. (2000). The effect of talk and writing on learning science: An exploratory study. *Science Education, 84*(5), 566–593.

Rupley, W., & Slough, S. (2010). Building prior knowledge and vocabulary in science in the intermediate grades: Creating hooks for learning. *Literacy Research and Instruction, 49*(2), 99–112.

Scruggs, T. E., & Mastropieri, M. A. (2013). PND at 25: Past, present, and future trends in summarizing single-subject research. *Remedial and Special Education, 34*(1), 9–19.

Scruggs, T. E., Mastropieri, M. A., Bakken, J. P., & Brigham, F. J. (1993). Reading versus doing: The relative effects of textbook-based and inquiry-oriented approaches to science learning in special education classrooms. *Journal of Special Education, 27*(1), 1–15.

Stahl, S. A., & Fairbanks, M. M. (1986). The effects of vocabulary instruction: A model-based meta-analysis. *Review of Educational Research, 56*(1), 72–110.

Taylor, D. B., Mraz, M., Nichols, W. D., Rickelman, R. J., & Wood, K. D. (2009). Using explicit instruction to promote vocabulary learning for struggling readers. *Reading & Writing Quarterly, 25*(2-3), 205–220.

Villanueva, M. G., Taylor, J., Therrien, W., & Hand, B. (2012). Science education for students with special needs. *Studies in Science Education, 48*(2), 187–215.

Wallace, C. S., Hand, B. B., & Prain, V. (2004). *Writing and learning in the science classroom* (Vol. Vol 23). Dordrecht, The Netherlands: Kluwer Academic Pub.

Yore, L. D., Hand, B. M., & Florence, M. K. (2004). Scientists' views of science, models of writing, and science writing practices. *Journal of Research in Science Teaching, 41*(4), 338–369.

Young, E. (2005). The language of science, the language of students: Bridging the gap with engaged learning vocabulary strategies. *Science Activities: Classroom Projects and Curriculum Ideas, 42*(2), 12–17.

Zywica, J., & Gomez, K. (2008). Annotating to support learning in the content areas: Teaching and learning science. *Journal of Adolescent & Adult Literacy, 52*(2), 155–165.

Chapter 6
Toward a Pedagogy of Cultural Relevance

6.1 Improving My Practice to Support the Science Learning of Sixth Grade African American Female Students

Mayra

I have a Ph.D. in pharmacology. When I began teaching, I was not necessarily the best at teaching sixth-graders. I was more of a scientist with a lot of content knowledge. I didn't have the educational courses for how to better teach my students. U-FUTuRES was a great opportunity. I had learned some things along the way because it was my third or fourth year of teaching, but it was a great opportunity to actually have the training in education. Even though I belong to a minority population (Hispanic), I did not feel prepared to face the challenges of teaching a diverse population of learners.

I teach at a laboratory school on a university campus, where the demographics of the student population is representative of the state. The student population is comprised of 48% White and the rest of the students are primarily African American and Hispanics. Approximately 20% of this diverse population is on free and reduced lunch. Like many people in our university town, I have a PhD with a high degree of confidence about my capabilities as a research scientist. In the professional development (PD) program, (U-FUTuRES), the online degree required the completion of a teacher inquiry in which I examine my own teaching. As a researcher, the process of collecting data and reflecting on my own work was at first very uncomfortable. I place a high value on numerical data collected and considered objective. But, never the less, I was interested in improving my teaching and understanding the possible effects on the science achievement of my 6th grade students. As a scientist now middle school teacher, I welcomed the outcome of effecting changes based on the specific needs in my classroom. Specifically, the program activities and the teacher inquiry in particular, gave me the opportunity to learn strategies to better support my students from populations underrepresented in science.

Many of the students entering 6th grade at my school are from elementary schools around the county and nearby districts, hence, they usually have different learning experiences in science. While some were exposed to inquiry-based science,

R. M. Pringle, *Researching Practitioner Inquiry as Professional Development*,
https://doi.org/10.1007/978-3-030-59550-0_6

others had traditional learning experiences including reading texts and watching science channels on the Internet. I teach a science curriculum that is consistent with the reform in K-12 science education which includes an inquiry-based approach to science teaching and learning. Therefore, in my class, students learn science by doing and in the process, they develop science knowledge and scientific and engineering practices (NRC, 2012). In teaching this curriculum, I have observed some positive results in student engagement and science learning but some students continue to be low achievers. I was clearly not meeting the learning needs of all my students as evident in their performances on both class and state assessments.

In the recent 8th grade state assessment, 20 out the 33 students performing below grade Level were female students. Eleven of the students were from underrepresented populations (8 African American, 2 Hispanic and 1 multiracial) of learners. I am not teaching 8th grade where the state tests are administered, but the students' 6th grade experiences are very important in preparing them for science through the middle grades. As the 6th grade science teacher, I want to provide the best practices to support all my students' learning. My wondering for this teacher inquiry project emerged from my wanting to explore the ways in which I can better support the science learning of the 6th grade African American girls. As I entered into the phase of wondering about the state of science achievement among the African American and Hispanic students, the following question emerged: How can I incorporate culturally relevant teaching strategies into an inquiry-based science curriculum? My focus on African American girls and culturally relevant pedagogy became critical, as I explored the literature and began to understand that lack of interest in science among African American females began at an early age (Pringle, Brkich, Adams, West-Olatunii, & Archer-Banks, 2012). My goal for this teacher inquiry was to investigate my practices in ways to support the science achievement of the African American females. To better support the development of my teacher inquiry, I set out to explore the literature on culturally relevant pedagogy.

6.1.1 Cultural Relevant Pedagogy in the Literature

As classrooms across America become more diverse, teachers are challenged to find ways to engage all learners in meaningful science learning. In order to allow access to science, educators have recommended a culturally relevant science program that teaches to and places value on students' lived experiences (Ladson-Billings, 1995; Wallace & Brand, 2012). The scholars suggest that in addition to traditional teaching strategies, teachers should design and use explicit strategies to encourage all students to consider themselves successful in science. They write that teachers who become culturally responsive understand the lives, cultural norms and values of their students. The teachers then use the knowledge the students bring to their lessons but also teach in ways that allow them to maintain their own cultural integrity (Ladson-Billings, 1995).

Culturally responsive pedagogy is divided into three functional dimensions: the institutional dimension, the personal dimension, and the instructional dimension (Richards, Brown, & Forde, 2007). According to Richards et al. (2007), the institutional dimension involves the administration's policies and beliefs about diversity; the personal dimension represents the teacher's own intellectual and emotional perception about diverse cultural backgrounds; and, the instructional dimension includes the various strategies teachers must use in their classroom to reach a diverse population of students.

According to educators, many different explicit strategies should be used to help all students be successful in science classrooms (Cox-Petersen, Melber, & Patchen, 2012; Loftin, Davis, & Hartin, 2010; Reyes, Brackett, Rivers, White, & Salovey, 2012). In her research Cox-Petersen et al. (2012) found that promoting inquiry-based instruction, diversifying seating and grouping, encouraging real, active, and engaged conversations, creating a safe environment and developing student's capacity to think and act independently are all strategies to promote culturally responsiveness in science teaching. As I wondered how and what strategies to use, I embraced the instructional domain of cultural relevant pedagogy and the notion that increasing student engagement is vital to science learning. As I moved into my teacher inquiry, I set out to examine ways to increase the response rate of the African American girls during the 6th grade Chemistry unit.

I selected four African American girls from three different teaching periods as the focus of my inquiry. Each girl was assigned a pseudonym and a number. Student #31 was a struggling student with an average reading level. She showed little consistency turning in homework assignments. I was the advisor for student #43 and developed a special relationship with her before the project. We met twice each week for 10 min during which we worked on organization and study skills. Even though her overall grades were low, she was usually motivated and asked questions as she participated in class. Student # 42 was very quiet in class and appeared disinterested in the science activities. Outside of class and along the corridor however, I usually observe her being very energetic and involved with her friends. Student #51 was a struggling reader with low achievement and was in a class with the highest academic achievers in the 6th grade.

6.1.2 Conducting the Inquiry

As I set out to explore how I could adjust my teaching practices to better support science learning among the African American girls, I collected both qualitative and quantitative data. I kept field notes and created a chart to document interactions during collaborative activities and also to measure their response rate. The indicators used to measure students' response rates were: students raising their hand, group discussions, answering the making sense questions and asking questions if they needed help. At the end of the day, the notes from the chart were summarized and transferred to my journal along with my reflections. I collected samples of the girls' class work including responses on tests and quizzes. In addition, I collected digital photographs of students' final projects and their graphic modeling activities.

Quantitative analysis included comparison of pre-post-assessment scores and student's first quarter, second quarter and first semester exam grades. Data analysis was ongoing during the inquiry. I read my field notes and added thoughts in my reflective journal and made judgments as to what practices were working and what needed to be changed as I planned for the next lesson. Several patterns emerged over time which lead to the identification of themes such as questioning, technology, seating arrangement, cooperative learning, communication, self-advocacy and confidence. I chose the most persistent themes which contributed to the generation of claims presented in the next sections.

6.1.3 Findings and Discussion

Some important patterns emerged as I focused on the instructional domain of culturally relevant pedagogy among four African American girls. Specifically, I investigated the use of appropriate culturally relevant strategies in increasing their response rate and achievement during the 6th grade IQWST Chemistry unit. With a focus on the instructional dimension, I examined dynamics within collaborative learning groups, responses to project work, and science achievement as indicated on the classroom assessment.

6.1.3.1 Strategic Grouping: African American Females Were More Engaged When Grouped Strategically

In organizing my classroom, I typically used deliberate and strategic seating arrangements with the intent to reduce classroom management issues. During the teacher inquiry and after careful observation and analysis, I found that the engagement in science lessons and the response rates of the African American girls were greatly affected by their seating arrangement. The girls were interactive at tables with mixed abilities and as they sat across from each other on square table arrangements. Student interactions varied when I changed seating assignments based on gender. For example, Martha, was typically quiet and rarely interacted. The following is a typical example of a journal entry, "Today, Martha was quiet, head down." I observed this pattern over a number of lesson periods. I realized that Martha was seated next to and partnered with a very confident and outspoken, high achieving white female student. I wondered if she would fare better if I partnered her with a less outspoken student. In the next lesson, her partner was an average achieving white male student. After the change was made, I observed that Martha's response rate and engagement increased. For example, my field notes stated, "Martha raised her hand at the beginning of the lesson, she completed all the notes, and asked me a question related to the class assignment."

Another pattern emerged as I observed Esther. She was seated at a table by an African American male student, who seemed to annoy her. I moved her, partnering her with a high achieving white female student. This "gifted" female student was shy and quiet but performed her work with high quality. Over time, Esther developed a relationship with the "gifted" female student. She began to ignore the annoying male student, who had continued to seek her attention. The gifted female student seemed to serve as a role model for Esther. After 4 weeks, Esther was engaged during classroom discourse and stayed on task to complete all her work before the end of class.

My overall analysis showed that strategic seating arrangements increased the engagement of each of my African American students. Depending on the interactions with their partners, African American students might gain confidence and participate more frequently in class. By being flexible in adjusting their seating arrangements, I had a positive response. This flexibility in the seating, focus on their interactions, and being responsive to the girls' needs resulted in remarkable improvement in the observed patterns of engagement in all areas of the science lessons.

6.1.3.2 Cooperative Learning (CL) Increased African American Females' Response Rates

The inquiry-based science necessitated much group work as the students developed scientific and engineering practices. Organizing the class into cooperative groups provided an environment that facilitated student interactions such as sharing, discussing, and argumentation as they complete assigned tasks. I used different approaches in organizing the cooperative groups and then observed the students' reactions. For example, during a lesson on mass and volume, Anna participated enthusiastically when paired with a group to work on vocabulary. Her team made flash cards and on one side they wrote the word or the term and on the other side they wrote the definition and or examples. This was followed by a partner quiz using the flash cards. At the beginning of the third lesson they quizzed their "shoulder" partner and during the next lesson, they quizzed the student sitting across from them. In conducting my formative assessment, Anna enthusiastically provided a definition of "mass." She not only raised her hand in response to oral questions posed to the class but also volunteered to go to the board to complete fill-in the blank tasks. Furthermore, on the unit quiz, Anna provided correct answers for the terms mass, volume and matter correct. Anna's increasing engagement with her peers during collaborative group activities also contributed to her learning during the unit.

In another approach to CL groups, students were placed in small groups as they generated claims based on evidence after they had conducted a science activity investigating the mass and volume of air. After they completed the measurement, students were instructed to respond to the following questions: Is air matter? What evidence can you provide to support your claim? Using the structure of Timed-Pair-Share, students discussed their answers and monitored the responses of their peers.

As they discussed, I walked around the room observing interactions and at times engaging in conversations with selected students. As I listened into Martha's group, she was explaining why she had said "No," in response to the question of air being matter. Martha provided her reasoning as she said, "I didn't think of air as an object." However, as she heard the other group members discussing their evidence, she changed her response. Martha's level of engagement increased during the interaction in the CL group. Her participation in the conversations in her CL group provided the evidence that challenged her misconception and lead to her understanding that air indeed had mass. CL not only increased the response rate of the students, but also provided a safe environment to share their work and learn with and from their peers.

Assigning class projects in which students developed models provided multiple opportunities for students to demonstrate their science learning within CL groups. In one lesson, students were asked to draw a model of air that would show its four characteristics. Each of my focus students worked with a partner from their CL groups. They stayed on task during the entire time allocated for the project (about 70 min). Esther and her partner drew a plan before actually making the model. Esther demonstrated an above level understanding of the concepts of compression and expansion as her model showed that the number of particles (mass) did not change during expansion and compression of gases. Anna worked quietly, but effectively, completing the project on time. She and her partner divided the work. One made a model that demonstrated adding and removing air; the other student modeled the expansion and compression of air. As I walked around checking their models, they were engaged in discussions and seemed confident talking about the development of their models. Martha, nearing completion of her model inquired if she could include her own examples into the project. While I was not surprised of her request, it certainly made me feel happy that she was extending her learning beyond the examples produced in the science lesson. Martha and her partner came after school to continue the work on their project and to create new and "better" models because they were not satisfied with their original work. Martha said she wanted to be creative and so she added a drawing showing air being pumped into a bicycle tire. As they turned in their assignment, Martha declared "this is the best" model. Overall, I was impressed with the response of the focus group students. This increase in engagement and demonstration of their science learning resulted in the achievement of passing grades on their projects.

6.1.3.3 Assessments Demonstrated Improved Academic Achievement

To determine the impact of the strategies on the African American girls' science learning, I made comparisons between pre and post responses to class assignments. I compared the grades for the first quarter (Q1) and the second quarter (Q2). With the exception of Anna, focus students received Bs in both quarters which can be interpreted as an above average level of performance. Esther showed the greatest increase in the second quarter (from 81% to 86%) as well as the greatest increase in engagement. I also compared the results of the pre- and post-unit assessments provided by the curriculum. The results showed that all four students had increases scores on the post-assessment, with scores increasing to an average of 80%. These results are evidence of the impact of using specific organizational strategies to encourage engagement and class participation.

6.1.4 Teacher Inquiry as Professional Development

At the beginning of the journey in the teacher inquiry project, I did not understand the difference between a reflective teacher and teacher inquiry. After all, "teachers reflect every day..." (Dana & Yendol-Hoppey, 2009). But, in a short period of time, I learned that the depth and intentionality of finding answers to my questions were the essence of teacher inquiry as compared with daily informal reflections. As I come to the end of my first teacher inquiry, I have realized that this is just the beginning of a lifelong educational journey. A journey in which I will constantly make adjustments to address observed issues of science teaching and learning among my students. I have come to embrace this act of critically reflecting on my practice as the essence of professional development as I seek to meet the learning needs of all my students.

In my teacher inquiry, I integrated into my teaching an aspect of the institutional dimension of culturally relevant pedagy (Richards et al., 2007). As I reflect on the process, I am aware that a focus on four girls out of a large population is not enough to drastically change the trajectory nor the persistence in marginalizations that exist in science teaching and learning. However, this is a start in addressing a specific problem of practice in my classroom. I was concerned about the science achievement of the African American girls. I set out to adjust my classroom arrangement to maximise their poisitiive engagement in my science lessons. The project resulted in an increase in their class participation as seen in the increased response rates of all the girls. I am still concerned however, about the extent to which the girls will reach the levels of achievement required throughout the middle grades and be better prepared for high school STEM related courses. As I prepare for my next cycle of inquiry, a number of questions have arisen as my wondering continues: How I can support the continued science learning of African American girls toward increasing their performance on state and classroom assessments? How can I adjust my teaching to include more personal dimensions of culturally responsive pedagogy?

6.2 Cultural Competence: It Matters in the Science Classroom!

Jennifer

I live in a very small and very poor rural county in Northern Florida. The County has only one middle school and one high school. The only major Civil War battle in Florida was fought in this county and is celebrated and re-enacted every year. The county is approximately 85% white. I grew up in this county and attended this school alongside some of my students' parents. Our African American students have struggled to be successful. There needs to be a change and this inquiry was the perfect opportunity to engage me in the process of moving toward change.

I am one of two White females assigned to teach 8th grade science in a small rural county. In this state, students in the 8th grade are tested annually in Science, Mathematics, and English Language Arts. Students' performances on these state tests are just one of the criteria for determining the annual school grade. For six consecutive years, our Middle School received an "A" grading from the state. Last school year, the school was downgraded to a "B" grade. The lowering of the school grade caused panic among the district administration and throughout our 8th grade teaching staff. We were therefore tasked to find the solution and ways to improve the achievement of all students so as to return to being an "A" rated school.

The school with an enrollment of over 1000 students, is housed in an old, red brick building that was once a high school. Approximately 85% of the students are white while the other 15% are minorities, predominately African Americans. A review of the students' science and mathematics scores on the state tests revealed an alarming trend. Even though the school was graded an "A" over the years, African American students had consistently scored below the state's average. Data from the past academic year showed 20 of the 23 African American students scored below the state average, and the other three performed just at the average level in science. Ironically, for years, we all knew some students were below average in their academic achievements across all subjects, but there seemed to have been a level of comfort with the overall academic achievement because of the school's rating. Among a group of 110, this year, I have 13 African American students. In the current climate at my school, and the need to address students' achievement, I have focused my attention on how best to meet the science learning needs of our low performing African American students.

6.2.1 *Moving into Action*

As a team, the 8th grade teachers began to have conversations about how best to support the African American students' overall academic performance in all subject areas. For the most part, the students were doing well in football and basketball and were making us proud among our rural neighbors. For my subject, the concern is how to teach the inquiry-based IQWST curriculum in ways that would improve the students' science achievement. One question that kept haunting me was: How do my African American students perceive science learning and my classroom? I thought if I could answer that question, I would have some guidance on how to proceed.

As I wondered about how best to address some of the issues raised and to fulfill the requirements of the degree program, I wanted to explore the use of culturally relevant teaching. Would intentional teaching strategies, including culturally relevant teaching in inquiry-based science increase the academic achievement of African American students? The literature on culturally relevant teaching provided some guidance as I began to explore how I could adjust my own cultural responsiveness in ways to complement the teaching of the curriculum.

6.2.2 Related Literature

Schools in the U.S. are culturally diverse places. In addition to diversity among student groups, teachers are also racially and ethnically different from their students (National Education Association (NEA), 2008). The differences in cultures between students and teachers have been identified as possible reasons why some groups of students are underperforming in academic achievement. A large number of teachers struggle to teach students from different cultures and the literature contains statistics highlighting the underachievement of children from minority populations especially African American children (Davis, 2007).

The achievement gap between African American students and their White counterparts in science is especially troubling. The academic achievement in middle schools determine the students' courses in high school and ultimately impacts their economic well-being as adults. In calling for quality education for African American to ensure success and preparedness for STEM-related careers, educators state that in addition to competence in the content of science and an engaging curriculum, teachers need to be culturally competent. When teachers are culturally competent they have the ability, skills, and knowledge to effectively teach students from diverse cultures (NEA, 2008). In addition, a teacher who is culturally competent understand the limitations reflected by his or her own sociocultural contexts and the differences with their diverse students (Barrera & Kramer, 1997; Craig, Hull, Haggart, & Perez-Selles, 2000). In order to improve the academic achievement of students of color, Gloria Ladson-Billings (1994) developed a theory of culturally relevant teaching committed to high levels of academic achievement, maintaining cultural integrity; and developing critical consciousness about the existing social order. Such teaching, rooted in social justice can address the iniquities in the educational experiences of students from historically marginalized communities (Allen, Hancock, Lewis, & Starker-Glass, 2017). In culturally relevant teaching, teachers use and validate the culture and life experiences of their students (Wallace & Brand, 2012). In so doing, teachers give credence to the students' cultural knowledge, prior experiences, frames of reference, and performance styles (Gay, 2010, 2018). The goal according to Wade Boykin (2002) and reported by Gay (2018) is to develop the talent potential of students of color and placing them at promise instead of risk.

Many students of color have not been achieving in schools as well as they should and for too long. In the literature and supported by research is the fact that to reduce the achievement gap between White and minority students, teachers will need to recognize, honor, and incorporate the personal abilities and cultural experiences of students by making changes to their teaching strategies. The educators maintain that teachers can use the students' experiences as points of references and motivational strategies to evoke the interest and involvement of minority students in their learning.

The literature is also clear about the positive impact of inquiry-based science learning in engaging minority learners. The features of the inquiry-based science curriculum require the full involvement of learners. The lessons are driven by learning goals and the students examine science phenomena, collect evidence, generate claims, reason, and provide explanation. Furthermore, each lesson begins with surfacing the students' prior knowledge about the learning goals and later in the lesson they would evaluate the extent to which they have confronted any misconceptions or have increased their understanding. The curriculum was well-designed to meet the learning needs of all learners. However, there was much concern about the achievement of African American students' as I teach this new science curriculum.

As I reflect on the heightened importance of culturally relevant teaching in the literature and the need to be effective with my students, I am now prepared to consider ways to adjust my teaching toward being culturally relevant. I am intrigued with examining how I can incorporate the students' cultural experiences in my science lessons. Many strategies have been identified but where do I begin? To better serve my students and offer them a classroom setting that they enjoy and hopefully excel in, I will need to begin to build positive relationships with the African American students and in so doing become knowledgeable of their experiences outside of the science classroom. The focus of my teacher inquiry became exploring ways to interact with the African American students and to value and include their experiences in my teaching. This also included the frequency and ways I would require them to share and participate.

6.2.3 Making Inquiry into My Teaching Practices

At the start of the teacher inquiry, the students answered 20 questions on a survey of grades 6–12 teachers. Their responses would help me to assess their perceptions of inquiry-based science and my teaching in particular. At the end of the 9-week period, the survey was again administered to all students. I collected artifacts of the students' work which included models they created in response to inquiry activities, assignments completed in class, grades on benchmark tests, lab grades, homework grades, and quizzes. For each science activity, I kept a tally chart of off and on-task conversations especially when the students chose their own partner as against when I made the choice. I also kept a journal in which I detailed as much as possible my observation of the students' responses to the culturally relevant teaching strategies. Each day as I wrote my description, I also reflected on each lesson. The students were in different periods and in my documentation, while I kept their identity during the data collection, my overall focus became the reactions and responses to the lessons. Data analysis began with tabulating the students' responses to the survey and identifying areas they identified as weaknesses in my teaching. During the process of the teacher inquiry, the journal was read a couple of times as I identified hunches. I strategically used some of the in-class hunches in conversation with the students especially when I visited their games. For example, after one science activity which required the students to work in pairs to develop a consensus model, a pair of African American boys wanted to know what I thought of their work. After school on the corridor, I inquired from them what of the lesson worked for them. One of the boys responded and among the ideas he shared, he felt, I was listening to him.

After the first 9 weeks of teaching and having administered the science test which was correlated to the district's pacing guide, I asked all my 8th grade students to once again complete the survey about my science teaching. I reviewed the surveys but focused mainly on the responses of the 13 African American students. The survey included 20 questions which required students to provide a yes, sometimes, or no response. The questions were related to characteristics of my teaching, such as (1) presents material in a variety of ways (hands-on, group, written, orally, etc.); (2) is approachable and willing to help me; (3) encourages and accepts different opinions; and (4) is involved and supportive of students within the school setting. All students said that the material was presented in a variety of ways (indicated by yes responses).

6.2.3.1 Developing Relationships Beyond the Classroom

While many White students indicated that I was approachable and willing to help them learn, only half of the African American students said yes at the start of the project. The other half said "sometimes." Questions 11 and 15 were related to whether they thought I was approachable and if I was fair and consistent but while the students enjoyed the hands-on activities they were ambivalent about my personal involvement and support. I realized I had some work to do with improving my relationships with the African American students. I became deliberate in my interactions with the African American students between classes, asked about their homework, and brought up topics that I knew interested them such as upcoming games. Several students played sports, so I started going to their basketball games and made sure I spoke to them after their games. The following is a conversation recorded in my field notes and which took place at a school basketball game.

Anthony: Hey Mrs. Richardson, who are you here watching?

JR: I came to watch you play. I heard how good you were and wanted to see it for myself. You played a great game.

Anthony: Thanks, but really who did you come to watch because the dance team didn't dance [I coach the middle school dance team and have attended games when they danced].

JR: I really came to just watch you play and I brought my husband and my sons too!

Anthony: I've never had a teacher come watch me play before.

Each time I read this encounter in my journal, it affirmed the fact that the African American students did not think teachers cared about them. As I reflected, it was clear that this was a testament to our actions toward them – real or imagined! I had been the first teacher to attend his games just to see him play, and I could tell it really meant a lot to him. The next day at school, Anthony told everyone in the class that I went to watch him play and that I brought my family with me. He made sure that I knew the dates of the home games, so I could come to watch him play.

Alyssa played basketball, so I attended a few of her games as well. At one of her games, I realized that I had graduated from high school with her mother. During the rest of the season, I deliberately sat with her mom during the games. After this game and noted in the journal, Alyssa clearly became more engaging in the science lessons, talked more in class and would ask a range of questions. Interestingly, her questions at times were about providing more explanations to help her understand. During one of the lessons, as I surfaced students' experiences, I noticed that the students would bring up our out-of-classroom experiences such as their sports' experiences related to the scientific phenomenon being studied. The following example is from my field notes (Lesson 3, Why Does an Object Start Moving?):

JR: Ok class what causes objects to start moving?

Alyssa: Like when I threw the basketball down the court and we won? The basketball started moving.

JR: Exactly, Like that! What did you do to the basketball to get it to move down court?

Alyssa: I pushed it. Really hard. Oh, I get it; I have to push something to make it move. Or I can pull it.

JR: You are on the right track! Alyssa asked a good question about pulling something. [I turned to the rest of the class.] Why don't we investigate her thought—everyone pull on the table and see what happens.

Alyssa: The table moves, so a push or a pull makes things move.

Alyssa made connections to the lesson based on her basketball game. From my observation, this was the first time she had connected an event in her daily life to a classroom discussion of scientific phenomena. By attending extra-curricular activities of my students, I could see a difference in my relationships with them and their willingness to relate personal events to classroom discussions.

6.2.3.2 Why Are All the Black Kids Over There? Strategic Grouping

As I was engaged in the teacher inquiry, I realized I was beginning to make other observations about the student groups at the school. For example, the African American kids during lunch break and afterschool were mostly together. In rare times, there may be a White student in these social groups. In my journal, I wondered why this was not obvious before and wondered how strategic grouping could be incorporated into my lessons. The science lessons usually include students working in small groups because I believed that in collaboration students develop scientific understanding. Most times, I would just require them to get into their groups. Usually the groups are formed along racial lines and in rare times an African American girl is in a group with White students. During the teacher inquiry and at the beginning of the physical science unit, I allowed the students to choose their own work partners. I believe that the choice would allow them to be more comfortable with whom they were working. Not surprisingly, students chose their friends as partners. Typically, an African American student would choose another African American student as a partner. I began to keep a tally chart during group work noting whether a pair of students were on-task or off-task. For example, during a lesson on forces in the 8th grade IQWST Physical Science unit, students were expected to discuss their background knowledge on forces. I allowed students to choose a partner and brainstorm ideas about the phenomenon they observed, discuss how we use forces, discuss why forces are important, and decide on some questions that they had about forces to add to our discussion. I circulated among the groups to listen to the conversations. I noted a high amount of off-task behavior during the brainstorming time. When looking at the data, I noted that when students chose their own partner or small groups, some students had more off-task behavior. The African American students had more frequent patterns of off-task behaviors when they chose their African American friends as partners. I described the following situations in my field notes:

> Anthony immediately began to discuss football with his partner as soon as I instructed the students to begin brainstorming. I reminded him to stay on task and told him to discuss the forces used in football. He was intrigued by my suggestion and began to get back on topic. As I circulated the room, most groups stayed on task discussing the phenomenon, how they used forces, why forces are important, and coming up with questions they had about forces. When I returned to Anthony's group, he had started a new discussion regarding football instead of forces, though he had written some force-related things down. His partner and friend, Alyssa, did not attempt to get him back on task nor did she share any ideas about forces at this time.

Brianna and Chloe jotted down the statement, "We use forces everyday" on their paper, but were talking about a family gathering that was coming up. I reminded them about the questions they were expected to discuss to get them back on task. As I moved to another group, Destiny immediately said that she didn't know anything about forces, so she didn't know what she could share. She chose to work with Caleb who is very talkative and without her input did most of the brainstorming. However, once Caleb had jotted down some ideas, he "wandered" to the group that consisted of Brandon and Chloe. I reminded Daniel five times about being off-task. Brandon and Chloe began on-task but that ended quickly because of Daniel who wanted to consistently talk to them. Choosing partners who were their friends seemed to result in more personal conversations and less discussion of the assigned topic, so I changed the way I organized composition of the groups.

Strategically Assigned Groups I assigned partners or organized the composition of small groups so that students worked with others from a different gender, race, and academic performance level. As students entered the classroom, I allowed them to sit in their familiar arrangements, but assigned partners during the experiential parts of the lesson. During seatwork, they could return to their familiar desks. With strategic grouping, I noted that African American students increased on-task behavior and active participation. Upon reflection, the intentionality, and communication of high expectations made a difference in the ways the students participated in the science activity. For example, in the lesson — "Which Forces Act on an Object?"— students had to analyze different apparatuses. Student partners analyzed each apparatus and answered questions about the forces involved in various motions. The activity took approximately forty-five minutes for the students to complete the four work stations. I circulated the room, listened to the conversations, and noted "off-task" and "on-task" conversations. I noted the following observations:

> Anthony was very quiet, waiting on instructions from his partner. When Anthony understood the directions, he began to manipulate the apparatus and stayed very involved with the task at hand. Anthony was briefly off-task as he transitioned to the next station, asking Alyssa if she was going to try-out for the basketball team this year. He was back on task when he arrived at the station. Most of the groups stayed on-task discussing the apparatuses. Alyssa began manipulating the apparatus at their station and asked her partner about the first question.
>
> Daniel who loves to "wander" around the room stayed focused and worked with his partner; he was off-task once during the activity, saying a quick hello to a friend walking by in the hallway.

When students were organized into work groups on the basis of gender, academic performance level, and race, the African American students were more likely to remain on task.

6.2.3.3 Questioning: The Link to the Outside Experiences

At the beginning of each lesson, IQWST provides questions prompting students to share their background knowledge related to the science investigation. I generally would pose the questions provided in the curriculum and wait for volunteers to share their thoughts. The African American students rarely volunteered. To encourage African American students to participate, I began to ask a question, wait, then use the strategy of pop-sticks to give equal opportunities to participate. Because of the number of African American students compared to White students, at times I had to be creative in student selection. As I observed the participation level, I would sometimes give students the choice of writing their responses. While this worked for all groups of students, in one of my classes, I learned that two African American students were good at art. During the wait-time, I began to give students the opportunities to represent their responses in art-form of their choosing or in writing. In that way, I told them, they would clarify their responses before sharing. This also provided me with a base from which to have later conversations with the students as the lesson developed. I would specifically refer to their responses during the later interactions. In my field notes, I noted that the participation of some African American students increased during these whole class discussions. In a conversation, I overheard the following conversations between two African American girls: "You better have your answer because she is going to come back and talk to you about it in the lesson." In my reflection, I noted, "Was this an indication that the students were recognizing that I was truly interested in their learning?" The intentional adjustments to the questioning strategy proved to be easy to implement and from my observation increased the level of engagement of the African American students in the class activities and the accompanying discussions.

I was surprised by the responses on the pre-survey but accepted that I needed to make personal connections with my African American students. I was unaware that the African American students did not feel supported in my classroom. My efforts to talk to them in the hallways and cafeteria and to attend their extracurricular events made a difference in my classroom. They appeared more comfortable; they spoke up and shared their experiences more often. Strategically grouping students and using wait time intentionally, clearly made a difference in on-task behavior and resulted in increased participation during cooperative learning. I also saw increased responsiveness to questions if I waited and called on specific students. I do not yet know, however, if students' performance on academic assessments will be increased sufficiently to impact the evaluation from the state. Nevertheless, I was pleased with the increased student engagement and participation.

My learning in this project has implications for my approach to my teaching and teacher inquiry after this PD program. I have recognized that teacher inquiry extends way beyond the teaching of one science unit. Heartened by the result of my first teacher inquiry, I will continue to focus on questions that emerge from the issues of science teaching in my classroom. My wonderings have not been laid to rest. I need to work harder to engage with the African American students as I seek to answer other questions that have emerged during this teacher inquiry. Now, I wonder if assigning specific leadership roles in the classroom would impact the quality of the African American students' engagement in science.

6.2.4 *Teacher Inquiry as Professional Development*

As I conducted my teacher inquiry, I gradually came to believe and appreciate how going through the process as discussed by Dana and Yendol-Hoppey (2009), untangled some of the complexities in my classrooms specifically with my minority students. Taking time to formally investigate my wonderings and confront the issue in my teaching, allowed me to gain more knowledge not only about my African American students but also about myself as a science teacher. By conducting this teacher inquiry, I had answers to my questions about the African American students in my 8th grade class and how I can be better at supporting their learning. Making and using something as simple as a tally chart in order to collect data lead me to discover how African American students stayed on tasks while working with different partners. Getting to know my students outside the classroom also made my African American students more involved in classroom discussions when they felt like they had something in common with me which in turn made me "more approachable" to my students. I became more opened to seeing my students as learners not only in the teaching times but outside of the walls of the classroom.

I think I had some success in this first teacher inquiry but my research questions have not fully been answered. I need to continue and go into more in depth to learn more about myself as a researcher. All the literature that was explored throughout this process discussed cultural awareness within classrooms and how all students can benefit from having a culturally-aware teacher. The teacher inquiry is now causing me to be explicit in my approach and wanting to know how to help my African American students succeed in middle school science. Based on what I have learned from this teacher inquiry, I am more apt to consider the cultural differences when planning my lessons.

6.3 Using Culturally Responsive Strategies to Increase Science Achievement Among a Group of 7th Grade African American Boys

Sara

I moved with my husband into this town as he began his residency at the university. After an unsuccessful job search which began before we relocated, I decided to apply for a teaching position. I passed the required tests for certification and opted to teach 7th grade science. My assignment was to teach science to students in the regular program of a school that also had a magnet program.

6.3.1 Introduction

I teach science to seventh grade students at a middle school in the northeast section of the state. My school has a magnet program but I teach science in the "regular" program. The students in the regular program are primarily from the local community. They were not selected for the academically enriched magnet program. The students did not fulfill the selection criteria such as academic achievements and had many disciplinary issues and school referrals in their elementary schools before middle schools. The racial composition of my 7th grade science classes is made up of mainly African Americans and a few White students. On the other hand, the student population in the magnet program of the school consists of mainly White students and children of immigrants whose parents are associated with the major university and other institutions in the town. The state's science test is not administered at the grade 7 level. A comparison of the 8th grade science results of the regular and the magnet program shows a large achievement gap between the two programs. The science scores of the students from the regular problem consistently fall below the average of the school district and that of the overall state. At the same time, research indicates that during the middle school years, in general, students' attitudes towards school become more negative, and self-esteem and perceived competence in school decline (Grolnick, Farkas, Sohmer, Michaels, & Valsiner, 2007). This decline is however more visible among populations of students who experience difficult social and economic conditions.

In my science classes, the group of students with the lowest level of engagement and achievement in science is the African American boys. They are usually disruptive during class activities and engage in much off-task behaviors, work at a slower pace, and usually class assignments are incomplete. In some cases, the assignments are not turned in for grading resulting in relatively low grades. In addition, at the end of the grading period, their test scores are low indicating that not much of anything was learned. This trend of low performance by African American students, according to researchers is continuing (Jacobson, Olsen, Rice, Sweetland, & Ralph, 2001; Phillips, Crouse, & Ralph, 1998). The researchers also indicate that the educational achievement of the Black males is compounded by the issues of higher disciplinary measures and they have a greater likelihood of being suspended and expelled (Kane, 2016; Kennedy, 2011). Furthermore, the research state that these students are more susceptible to criminal behaviors and substance abuse leading to high incidences of being high school dropouts.

The phenomenon I was encountering with the low achievement in science by the African American boys was not unique to this school or program. In the literature, when compared to their White peers, even middle class African American males were lagging significantly behind their peers as indicated on standardized tests (Noguera, 2003; Wyatt, 2009). While many argue about the efficacy of these tests, as a middle school science teacher, I was mindful of the long-term impact on the lives of the students beginning with their high school placements. Furthermore, the students' achievements on these tests determine to a large extent the school's rating and the recent teacher accountability measures that were introduced.

During the PD program, I was introduced to the idea of equity in teaching, defined by Gorski (2013) as the "skills and disposition that enable us to recognize, respond to, and redress conditions that deny some students access to the educational opportunities" (p.19). Now that I have recognized the issue, how can I adjust my teaching to better accommodate the science learning of African American boys? My wondering question therefore becomes: What culturally relevant teaching strategies can I employ in my teaching to enhance the science learning of the African American boys in my 7th grade?

6.3.2 Culturally Responsive Practices

Many educators who have conducted research on performance in schools have noted that students, especially those that live in poverty experience many inequities when compared to their wealthy counterparts, face many challenges that impact their view of school and their engagement and learning in general. These researchers have concluded that there needs to be a shift in teaching strategies to make them more culturally responsive to the learning needs of all students (Gay, 2010; Ladson-Billings, 1995), especially those who are underrepresented in subjects such as science. This shift begins with creating and maintaining an equitable classroom environment. That is an environment in which students irrespective of ethnicity, gender, experiences or economic situation is appreciated, respected, feels welcome and is provided for equitably as any other member of the class.

In the article, *A Resource for Equitable Classroom Practices,* the author quoting Gay, described culturally responsive teaching as "using cultural knowledge, prior experiences, frames of reference, and performance styles of ethnically diverse students to make learning encounters more relevant and effective for them," (Montgomery Public School, 2010; page 29). Gorski (2013), believes the success of establishing and maintaining an equitable classroom relies on teachers understanding of equity and social justice and the role of culture in teaching. According to Swalwell (2011), teachers need to have an appreciation of diversity and develop skills and abilities to teach all students regardless of culture. These educators are clearly communicating that an understanding of cultural differences is important in teachers' practices as they teach in diverse classrooms.

As I teach science to 7th grade students, I often experience "on and off engagement" with some of my students many of whom have been identified as living in poverty. There are days when they enter the classroom with bright smiles on their faces, lots of energy, prepared and ready to participate in the lesson activities. On other days they appear lackluster and unwilling to collaborate with other students or participate in the lesson. It is not uncommon to hear these students say that they did not sleep well last night or the pace of the lesson is too fast or other students are making fun of them or that teachers do not understand or care about them. These students however, do not readily share their experiences with the teacher and usually when other students learn of situations these students are experiencing they become embarrassed and withdrawn. Toward achieving equity, teachers must have a heightened awareness of the situations of all learners and to become more sensitive to the needs of children who are from areas of high poverty. Gorski and Swalwell (2015), suggest that teachers cultivate the abilities to establish and maintain bias-free and discrimination-free classrooms, which requires an understanding and appreciation of the diverse cultures in the classrooms.

As I continue to identify strategies to make my practice more culturally responsive while increasing students' engagement and science achievement, I am drawn to the *warm demander no excuse stance*, posited by Bondy and Ross (2008). In the warm demander stance, teachers build a caring relationship for students while maintaining an authoritative presence in the classroom. When teachers are guided by this theory, they show care for their students' learning and well-being, seek to convince the students they can consistently produce good work, hold the students to high expectations, and create a safe and supportive environment for learning while maintaining good discipline.

6.3.3 Inquiring into Teaching Science to African American Boys

Once I decided on the subject area of my teacher inquiry, I reviewed the grades and scores of the eight African American boys who were in my first period science class. I noted the number of missing assignments and differentiated the test grades from the homework. My first data source was the boys' grades that were achieved in the first term of the school year. During the teacher inquiry, I created a performance chart that showed the students' grades, missing and late assignments and a column for notes. After the consents were received from the parents for them to participate in my teacher inquiry, I shared the chart with the boys and told them I was going to keep notes on their actions during my science lessons and our afterschool sessions. These sessions were organized to provide additional help with their science assignments. These notes were purely descriptive and were separate from my journal entries. My planning was second period and that gave me a brief time to write and reflect in my journal.

For data analysis, I read my journal entries and used colored highlighters to identify specific moments in my teaching and interactions with the boys who were involved in the teacher inquiry. I used yellow highlights when there were negative; red showed my frustrations; and green showed moments of cooperation by any or all of the boys. The coloring helped me to readily identify trends and patterns and would guide any changes that I needed to make in my teaching. For example, I realized my frustrations became heightened when the boys organized into two groups were not only off task but were causing other groups to become involved in off-task behaviors. Further observation showed that the boys' off-tasks behaviors were triggered when there were obvious issues with completing the activity in the lessons. For example, one lesson required the measurement and calculation of densities of cubes made of different materials. As issues around their understanding of the activity surfaced, the boys became engaged in playing with the materials. This led to the group not completing the task which included calculation and the generation of claims based on their observations.

My ongoing analysis of my data was extremely revealing about my view of the African American boys, my expectations, their responses to science learning and their perceptions of schooling. All eight boys who were the focus of my teacher inquiry were African Americans. They were similar in socio-economic conditions but I quickly realized they were eight different personalities. The teacher inquiry challenged my approach to science teaching and my expectations for groups of students – specifically the African American boys. As I worked closely with the boys, a trusting relationship and a safe learning space had to be created into which culturally relevant teaching was realized. This did not happen easily.

6.3.3.1 Teacher Expectations and Students' Experiences: A Mismatch

My analysis of the students' work before the start of the teacher inquiry showed extremely low performances in all three areas – test and quizzes, homework, and classwork. All grades for all the boys were unsatisfactory and a common trend was missing homework and incomplete classwork. At the start of the project, I had an afterschool meeting with the boys. Because I wanted to have the meeting with all eight boys together there were three cancellations. The class knew I was in a program at the university. During the meeting, I told the boys I wanted their help to complete an assignment in one of my classes. I reminded them of the script that was read and signed by their parent or guardian. While they did not express a discomfort, they wanted to know what exactly they were going to do. In my journal of that meeting, I realized I was not clear in describing what I was requiring of them during the period. "What would I want them to do?" I wrote. I had to think hard because at the meeting, I felt I did not provide them with a viable response. I wanted to know how best to reach and teach them science but it was difficult to communicate that I wanted them to respond in ways that would help me to achieve my goal. After the first introductory meeting, I tried to be clear on what my expectations were of the eight African American boys in my science class. After this meeting and reflecting on my expectations, I wrote in my journal: "I expect the boys to learn science as I magically (smile) transform my teaching to meet their learning needs."

I did not ask the boys about their expectations of the science classes. It was clear they felt the lessons were not about them but about the completion of my course work. In one of my lessons when I sat with one of the groups of boys working on a collaborative activity, Caleb asked if I was getting my work done. I asked him what he meant and as he repeated his question, Chris was quick to respond in a half serious manner, "are we helping you to finish your work?" All the boys looked around at each other as if I were not sitting there with them. The boys were experiencing the science lessons differently from what I was expecting. Their perception was that my university course work was important and their role was to do what I wanted them to do so that I could be successful. There was a clear misunderstanding. They were experiencing my science lessons but differently from the expectations of the teachers and the school system.

6.3.3.2 A New Twist on Engaging Students' Prior Knowledge

One of the strategies recommended in the literature on CRT is to incorporate students' out of the classroom experiences in the lessons. This is also the approach in the reform curriculum being used – to surface the students' prior knowledge which later in the lesson is used in the process of sense making. I tried to begin each lesson as indicated in the curriculum. Usually, I share the learning goals for the lesson and would invite the students to share their knowledge. This was always a low point in my lessons as very rarely the students who responded were able to make connections to the learning goals. As I began my teacher inquiry, I was therefore mindful of engaging the students lived experiences to support the science learning. The boys however were not responding to the whole class activities, so I decided to use this strategy of surfacing the ideas when they were working in their small groups – I had a captured audience of four in each group. During these sessions, most of the times they were eager to share, reacted to their peers, and at times even poked fun at each other. During these small group interactions, as I solicited the students' prior knowledge, I would write their responses on a mini white board. The mini white boards were typically used by the students during group activity to write their group consensus. I suggested to the boys that going forward they could take turns to write on the whiteboard. This suggestion was not easily embraced and in one group, the other three boys suggested that Ty should do the writing. Of the eight boys, Ty was usually reluctant to share and was mostly very quiet in class and also during the lunch period. However, Ty is often observed as being the instigator of disruptive behaviors and then would quickly appear to be uninvolved. He was also the lowest achiever in terms of academics and after 3 weeks, still had not submitted an independent work. In one of the small group conversations, as I tried to engage the group about acceleration and motion, Jason asked Ty why he was afraid to talk when he had a skateboard that he rode to school. Ty's response was that Jason's observation about his willingness to share was not true.

6.3.3.3 Building Relationship Before Pressing for Product

I thought long and hard about my work with the boys during the small group activities and what was being accomplished in terms of their science learning. I wanted them to feel relaxed, comfortable and to be fully aware that I genuinely cared about their science learning. I set out to build a caring and understanding relationship. I was guided by Hall and Hall (2003) who stated that one should consider strategies such as gentle intervention, avoid punishments and ensure teaching activities that ensured success for all students when trying to build positive relationships. Rather than the usual focus on their missing work, I invited them to talk openly about what was working for them in the science class. Specifically, I engaged them in casual conversations about sports and gently lead the discussion toward science. In one meeting, I asked the boys if they were the science teacher what would they do differently from what I had been doing over the past weeks. I prompted Ty a little and he stated that he liked the science activities but he did not like to write and neither did he like being called out in class. Jason and John also did not like being asked a question when they did not know the answer. John said he would be embarrassed if he gave the wrong answer. What was telling was that all eight boys liked the idea that during the science lessons, I took time to work with them in their groups. They felt special!

In one of our outside conversations, Ty finally explained that he preferred to work with others in the class and identified two African American girls with whom he felt comfortable. Even though he was not achieving in science, he said the girls would be more supportive and would listen to his suggestions which would also help him to learn better. According to Brackett, Reyes, Rivers, Elbertson, and Salovey (2011), students who are members of an emotionally supportive classroom will demonstrate better attitude towards school; higher achievement, greater engagement, enjoyment and motivation. *In addition,* Patrick, Ryan, and Kaplan *(*2007*) emphasized the need for teachers to organize their classroom environment for students to be engaged in their learning, to share their thoughts and feel safe to take risks. Research has also supported the idea that when students' suggestions are incorporated into the planning and implementation of activities to support their learning, they are more likely to "buy in" and generally lead to desirable learning outcomes. With this in mind, I moved Ty and placed him in another group as he requested.* The silence that normally surrounded Ty's participation in the small group with the other African American boys was replaced with conversations about football or basketball. Ty started showing increased class engagement and was sharing more in the group activities with the girls. On one occasion, Ty's group was randomly selected to share their procedure in making and testing the speed of a balloon tied to a string. The girls in Ty's group selected him to report and could be heard saying, "Ty you can do it" and he responded in the affirmative. The entire class cheered and he gave a wide smile on his way back to his seat.

According to Weinstein and Novodvorsky (2015), communities develop only when teachers provide opportunities for students to learn together in supportive ways and share experiences. As the project developed, I was moved by the boy's commitments to working together on the science activities. I wondered if they were genuinely participating for their learning or just doing it for my project. After a few weeks of increased concerted effort on my part to build a positive relationship with the boys, I noticed incremental positive changes in their overall conduct and attitude in science class. On occasions, at least five of the boys were first to the classroom even though at times they remained outside the door. I started to verbally welcome them into the classroom and to the science class. Later, I assigned them the task of extending a welcome to the other students as they entered the room. While there were definite changes in their deportment as learners in the science lesson and in their participation and getting more comfortable as learners, I have not observed much progress in the quality of their science learning. I know this is early in the process. I have made some adjustments to my approach and teaching these students. I have built a good relationship now I need to explore how I can translate my interactions into science learning gains.

6.3.4 *Learning About My Teaching: The Importance of Practitioner Inquiry*

One thing I have learned as I worked on being culturally responsive with a focus on the African American boys is that I had to confront and replace my own expectations and perceptions of my students. The challenges these boys face as learners in my science class needed to be addressed as they were low performing and showed little interest in learning science. As I set out to conduct the teacher inquiry, I was hopeful but wondered if the disparities in achievement could be addressed in the transformation of my teaching. As I developed positive relationships with them I had to focus on their potentials along with understanding their social experiences. I realized over the period that I was unaware of the underlying causes of their under achievement and lack of engagement in science. The boys' perceptions of teachers were that they were uncaring and schools were hostile places that they had to go to on Mondays to Fridays. Hostilities were directed to them as they were constantly being berated about their grades and their lack of producing the "products." They however were in an environment that was not catering to their academic needs and did not value the experiences they brought to school. As I reflected on my interactions with the boys, their actions constantly reminded me of their uniqueness and that "one size does not fit all."

The teacher inquiry has started me on my way to being more reflective and proactive in my teaching and in responding to the African American boys in my classes. I continue to improve on the process of building positive relationships and implementing strategies that engage them in meaningful ways.

The course readings and the teacher inquiry have changed my mindset as I reflect on my role as a science teacher. It is not only about the science knowledge and skills. I need to be responsive to the needs of all learners and to take a step back from historical expectations. The development of a positive relationship with the African American boys occurred over time. The boys felt misunderstood and undervalued in schools (Emdin, 2011) and experienced schools in ways different from the expectations of the teachers and the school system. Researchers have expressed that if students are convinced that their teachers care for them and believes in them, their learning and performance will increase. The boys became engaged in doing the science activities as they recognized that the relationship I was building with them was not only for my course project but was genuine toward their learning of science. This was also evident to them as I committed time to working with them and made visual changes in the lessons to accommodate their participation. The changes I experienced with the eight boys were due to my deliberate attempt at creating a learning environment that allowed them to feel safe as learners. It allowed them to move out of their levels of comfort and took risks, which upon reflection, should have been the norm in their schooling experience and not an effort in time.

6.3.5 Final Thoughts and Further Questions

The students in my classrooms are predominantly African Americans. In my first teacher inquiry, I explored ways to encourage and support their science learning. The rapport I developed with the students offered them spaces in which they became active participants in the classroom. Now, I wonder about continuing the process and making the necessary adjustments to my teaching. I am sure much is left to be done. Having started on this first teacher inquiry, I wonder about the ways I need to continue so as to impact the trajectory of the African American boys leaving my science class for 8th grade. Will I be able to continue the intense focus on their science learning? If the support is only being offered in their science classroom, what will be the overall impact? My learning in this project has opened my eyes that teaching is complex but teacher inquiry has the potential to allow teachers to work through problems of practice one issue at a time.

6.3.6 Researching Practitioner Inquiry: Developing Cultural Competence

A pressing challenge facing the education system is the persistent academic achievement gaps that exist along racial, ethnic, and socioeconomic lines. Even amidst current reform efforts, standards-based education movements, increased testing, and an unprecedented surge in charter schools, there is the failure to reduce the science achievement gap among populations of learners. The subject of the teachers' narratives in chapter six, emerged in response to the observed differences in academic achievement among diverse students in their middle school science classrooms. All three female teachers, two White and one Hispanic, acknowledged their struggles to engage and facilitate substantive science achievement especially among their diverse students. Each teacher began her practitioner inquiry project by exploring the plethora of literature on culturally responsive educational practices. They adapted relevant strategies in direct response to the problem of practice observed in their own classrooms, stepped out of their comfort zones, and in the process, reported varying levels of success. The pedagogical adjustments included attention to awareness and appreciation of the students' culture, enactment of culturally responsive management practices, and a focus on positive teacher-student and student-student relationships.

6.3.6.1 Cultural Responsiveness in Science Teaching

In a diverse classroom, the teacher and some students come from different cultural backgrounds. As schools in the U.S. undergo a dramatic change in demographics, the cultural gap between teachers and their students have been identified as one of the factors in students' academic performance (NEA, 2008). In the quest to better meet the needs of diverse learners and reduce the achievement gap, extensive studies have focused on the interactions among culture, learning, and academic performance. The discussions of the impact of culture on students' achievement, began from the premise that education must be specifically designed to support students' cultural integrity as well as academic excellence. Hence, attention to culture, according to Gay (2013), should enrich the culture of the diverse learners and equip them with the tools to become scientists if they so choose and all, as citizens, understand the enterprise of science. Teachers who are culturally competent display personal and interpersonal awareness and sensitivities to students from cultures other than their own (Diller & Moule, 2004).

In their practitioner inquiry, teachers examined and tailored aspect of their practices to particular groups of students. Consistent with the suggested practices in the educational literature, the teachers set out to incorporate culturally responsive teaching practices. Scholars have given credence to the viability of engaging cultural experiences and linking students' knowledge, skills, and experiences to curricular goals (González, Moll, & Amanti, 2005; Lee, Quinn, & Valdés, 2013). In the enactment of curriculum, teachers make pedagogical choices that can either support or constrain students' science learning (Pringle et al., 2012). When teaching is culturally responsive, students' cultural integrity as well as academic excellence are both facilitated and enhanced regardless of differences between the teacher and the students.

A definitive area in the teachers' narratives was their maintenance of high cognitive demand of all students and engaging them in the rigor of science learning tasks. At times, teachers exercised flexibility in their requirements without lowering the expectations. For example, all students selected ways of making their thinking visible. The students skilled in the arts were allowed to create representations of their ideas or thinking. Then, they shared their thinking and with teacher and peer support made connections to the new knowledge. These representations also became the central focus during the small group activities that accommodated more deliberate interactions. During the small group settings, teachers modeled expectations, acknowledged where appropriate the experiences shared by the diverse students, and provided additional time and scaffolding as needed for completing tasks. Activities in these small groups encouraged greater student engagement, increased students' level of comfort in sharing their experiences, and became the site for much collaboration on learning tasks. Teacher's focus on collaborative grouping strategies supported the social context of learning and allowed for the development of classroom relationships that valued the input of the diverse students. Here the voices of the students were more likely to be valued and the learning context shifted from a teacher-centered to a student-centered model (Cooper & Robinson, 1998).

Learning is enhanced when attention is paid to the knowledge and beliefs that learners bring to the learning task and used as a starting point for new instruction (NRC, 2000). Surfacing prior knowledge during instruction confronts and challenges misconceptions, create learning spaces for students to construct meaningful science learning, and support the monitoring of students' changing conceptions. Though aware of the cultural heritage, the teachers recognized they needed to learn more about their students. They therefore set out to build relationships outside of the formal classroom settings. For example, teachers interacted with students in informal conversations during lunch break and showed interest in students' extra-curricular activities. As they learned more about their students they were able to frame and pose metacognitive questions and design relevant metaphors and images that connected students' prior knowledge and the new concepts to be learned. In addition, these prior knowledge constructions, garnered from the students' lived experiences, provided further insights into how they made sense of their experiences.

Cultural responsiveness is embraced as more than just a set of practices embedded in curriculum and instructional units (Howard & Terry Sr, 2011). Its theoretical roots are embedded in the notion that learning, a socially mediated process is related to students' cultural experiences (Irvine, 2009). The teachers recognized the rich and varied experiences that diverse students brought to the learning environment. These experiences when used as frames of reference, build on the premise that while learning may differ across cultures, success can be achieved by valuing and validating students' cultural backgrounds such as their inclusion into instructional practices (Allen et al., 2017; Gay, 2013; Irvine, 2009; Ladson-Billings, 1995). Incorporating students' experiences in instructional activities is easier when teachers and students share the same cultural heritage – this was not lost on the teachers. In these learning environments, teachers recognize and understand the nuances of context in the students' actions including behavioural clues, comments, and questions (Delpit, 2006) and with ease, are able to elicit and connect students' prior knowledge. However, culturally competent teachers interact positively with the lived experiences of their diverse learners and make the link between the students' culture and the new knowledge and skills they encounter inside school.

The teachers were aware of the challenges inherent in being responsive to the sociocultural and linguistic backgrounds in their diverse classrooms. Responding to the challenges, they examined and adjusted their classroom practices. They embraced flexible grouping strategies, attended to the use of metaphors and images that connected to students' prior knowledge and the science concepts being learned. The expectation is that all teachers will become aware of the role of culture in their teaching and will seek to develop knowledge, skills and sensitivities to the diverse cultural experiences in their classroom. This is the essence of cultural responsiveness in science teaching.

6.3.6.2 Culturally Responsive Management Practice

Having a repertoire of effective classroom management strategies is an integral part of teachers' routine and daily practices. Primarily, these strategies seek to restrict off-task behaviors regarded as limitations to teaching and student learning. The behavioral norms in the classrooms, are established by teachers, who as the authority make their expectations clear to students. These expectations exist alongside strict adherence to rules and accompanying levels of punishments for behaviors deemed as disruptive (Bondy, Ross, Gallingane, & Hambacher, 2007; Woolfolk-Hoy & Weinstein, 2006). The decisions concerning acceptable behaviors are influenced by teachers' perceptions, values, and expectations. Thus, as members of the dominant culture, classroom management decisions are consistent with mainstream values and in effect, ignore the range of student diversities (Emmer & Stough, 2001; McCarthy & Banally, 2003). In schools with high population of students from marginalized populations, many of the decisions made by teachers are driven by the need to manage students' behaviors (Kennedy, Brinegar, Hurd, & Harrison, 2016; Traynor, 2003).

An abundance of literature highlights the misinterpretations and unnecessary disciplinary interventions that have occurred when teachers and students come from different cultural backgrounds (Kennedy-Lewis, 2013; Weinstein, Tomlinson-Clarke, & Curran, 2004). Consistent with the literature, the teachers recognized the connections between their students' engagement and the management system established in their classrooms. As the teachers questioned their traditional assumptions of "what works" in classroom management they became alerted to the tensions between conventional management strategies and the cultural backgrounds of their students (Ballenger, 1999; Brackett et al., 2011). Teachers acknowledged the differences in participatory patterns among groups of students and instituted strategies that maintained a focus on science learning. Some of these strategies included redirecting and maintaining the focus on science learning, using predetermined cues to encourage on task actions, and communicating high expectations to students. The success of these actions was further elevated as connections between the classroom management strategies and the development of positive teacher-student and student-student relationships was established. The focus therefore moved from the taken for granted practices of a management system grounded in dominant values and beliefs to the recognition of the cultural disconnect and its impact on the climate of the classroom. Ultimately, the goal of classroom management is provision of a learning environment that challenges learners to perform to their full capabilities. During instructional activities, through their words and actions teachers not only communicate the importance of learning but also build caring and supportive relationships (Bondy et al., 2007).

Teachers are more likely to create risk-free learning environments that support, nurture and respect students when relationship building is intertwined into culturally responsive classroom management. In classrooms in which culturally responsive classroom management is embraced, teachers further the development of positive relationships outside of the formal science classrooms. Teachers attend afterschool games, interact with family members, and in general learn about the students' cultures. Deliberate attempts are also made to understand the peer grouping dynamics of diverse students and their interactions in places such as the cafeteria and other meeting areas within the campus. These relationships built outside the classrooms become very impactful as teachers create contexts that respect and value students' cultural heritage. With such interactions, students are more likely to cooperate with teachers and in the process, some of the negative messages and experiences of the students become countered (Cothran, Kulinna, & Garrahy, 2003; Weinstein & Novodvorsky, 2015).

As teachers affirmed the students' personal and cultural identity, noticeable changes in student dispositions were observed. Teachers' observations were consistent with Brackett et al. (2011), that once the students felt they were being supported, they demonstrated better attitude towards learning and moved toward higher achievement and greater engagement in class activities. For example, students made deliberate attempts to be early to class, off-task activities were reduced, and they willingly sought out assistance beyond the formal classrooms in an effort to complete science assignments. Regardless of race, gender, class, ethnic, and SES, classroom environments should embody instructional practices and teacher-student interactions and relationships that foster students' motivation and engagement (Spearman & Watt, 2013).

Researchers examining the impact of conventional classroom management strategies and the punitive effects on some learners, posit that these environments are not supportive of student engagement nor do they minimize task avoidance (Patrick et al. 2007; Weinstein et al., 2004; Bondy et al., 2007). They offer that diverse students fare better in environments that attend to culturally responsive classroom management. When classroom management is employed from a position of the cultural heritage of students, their behaviors are not interpreted through the lenses of mainstream culture. That is, teachers aware of their implicit and explicit cultural biases will be more likely to shift the focus of management from retribution, to sites of safe haven where students' funds of knowledge are valued and celebrated and used as foundation for further learning.

References

Allen, A., Hancock, S. D., Lewis, C., & Starker-Glass, T. (2017). Mapping cultural relevant pedagogy into teacher education programs: A critical framework. *Teachers College Record, 119*(1), 1–26.

Ballenger, C. (1999). *Teaching other people's children: Literacy and learning in a bilingual classroom*. New York, NY: Teachers College Press.

Barrera, I., & Kramer, L. (1997). From monologues to skilled dialogues: Teaching the process of crafting culturally competent early childhood environments. In *Reforming personnel preparation in early intervention: Issues, models and practical strategies* (pp. 217–251). Baltimore, MD: Paul H. Brookes Pub. Co.

Bondy, E., & Ross, D. D. (2008). The teacher as warm demander. *Educational Leadership, 66*(1), 54–58.

Bondy, E., Ross, D. D., Gallingane, C., & Hambacher, E. (2007). Creating environments of success and resilience: Culturally responsive classroom management and more. *Urban Education, 42*(4), 326–348.

Boykin, A. W. (2002). The triple quandary and the schooling of Afro-American children. In U. Neisser (Ed.), *The school achievement of minority children: New perspectives* (pp. 57–92). Hillside, NJ: Erlbaum.

Brackett, M. A., Reyes, M. R., Rivers, S. E., Elbertson, N. A., & Salovey, P. (2011). Classroom emotional climate, teacher affiliation, and student conduct. *The Journal of Classroom Interaction, 46*, 27–36.

Cooper, J., & Robinson, P. (1998). Small group instruction in science, mathematics, engineering, and technology. *Journal of College Science Teaching, 27*, 383.

Cothran, D. J., Kulinna, P. H., & Garrahy, D. A. (2003). "This is kind of giving a secret away...": students' perspectives on effective class management. *Teaching and Teacher Education, 19*(4), 435–444.

Cox-Petersen, A., Melber, L. M., & Patchen, T. (2012). *Teaching science to culturally and linguistically diverse elementary students*. Boston, MA: Pearson.

Craig, S., Hull, K., Haggart, A. G., & Perez-Selles, M. (2000). Promoting cultural competence through teacher assistance teams. *Teaching Exceptional Children, 32*(3), 6–12. Curriculum Inquiry 43:1 (2013).

Dana, N. F., & Yendol-Hoppey, D. (2009). *Facilitator's guide: The reflective educator's guide to classroom research: Learning to teach and teaching to learn through practitioner inquiry*. Thousand Oaks, CA: Corwin Press.

Davis, P. E. (2007). Something every teacher and counselor needs to know about African-American children. *Multicultural Education, 15*(3), 30–34.

Delpit, L. (2006). Lessons from teachers. *Journal of Teacher Education, 57*(3), 220–231.

Diller, J. V., & Moule, J. (2004). *Cultural competence: A primer for educators*. Toronto, ON: Thomson Wadsworth.

Emdin, C. (2011). Moving beyond the boat without a paddle: Reality pedagogy, Black youth, and urban science education. *The Journal of Negro Education, 80*, 284–295.

Emmer, E. T., & Stough, L. M. (2001). Classroom management: A critical part of educational psychology, with implications for teacher education. *Educational Psychologist, 36*(2), 103–112.

Gay, G. (2010). *Culturally responsive teaching: Theory, research, and practice* (2nd ed.). New York, NY: Teachers College Press.

Gay, G. (2013). Teaching to and through cultural diversity. *Curriculum Inquiry, 43*(1), 48–70.

Gay, G. (2018). *Culturally responsive teaching: Theory, research, and practice*. New York, NY: Teachers College Press.

González, N., Moll, L. C., & Amanti, C. (2005). *Funds of knowledge: Theorizing practices in households, communities, and classrooms*. New York, NY: Routledge.

Gorski, P. (2013). *Reaching and teaching students in poverty: Strategies for erasing the opportunity gap*. New York, NY: Teachers College Press.

Gorski, P., & Swalwell, K. (2015). Equity literacy for all. *Educational Leadership, 72*, 34–40.

Grolnick, W. S., Farkas, M. S., Sohmer, R., Michaels, S., & Valsiner, J. (2007). Facilitating motivation in young adolescents: Effects of an after-school program. *Journal of Applied Developmental Psychology, 28*, 332–344.

Hall, P. S., & Hall, N. D. (2003). Building relationships with challenging children. *Educational Leadership, 61*(1), 60–63.

Howard, T., & Terry, C. L., Sr. (2011). Culturally responsive pedagogy for African American students: Promising programs and practices for enhanced academic performance. *Teaching Education, 22*(4), 345–362.

Hoy, A. W., & Weinstein, C. S. (2006). Student and teacher perspectives on classroom management. In C. M. Evertson & C. S. Weinstein (Eds.), *Handbook of classroom management: Research, practice, and contemporary issues* (pp. 181–219). Lawrence Erlbaum Associates Publishers.

Irvine, J. J. (2009). Relevant: Beyond the basics. *Teaching Tolerance, 36*, 41–44.

Jacobson, J., Olsen, C., Rice, J. K., Sweetland, S., & Ralph, J. (2001). Educational achievement and Black-White inequality. *Education Statistics Quarterly, 3*(3), 105–113.

Kane, J. M. (2016). Young African American boys narrating identities in science. *Journal of Research in Science Teaching, 53*(1), 95–118.

Kennedy, B. L. (2011). Teaching disaffected middle school students: How classroom dynamics shape students' experiences. *Middle School Journal, 42*(4), 32–42.

Kennedy-Lewis, B. L. (2013). Persistently disciplined urban students' experiences of the middle school transition and "getting in trouble." *Middle Grades Research Journal, 8*(3), 99–116.

Kennedy, B. L., Brinegar, K., Hurd, E., & Harrison, L. (2016). Synthesizing middle grades research on cultural responsiveness: The importance of a shared conceptual framework. *Middle Grades Review, 2*(3), 1–19.

Ladson-Billings, G. (1994). *The dreamkeepers: Successful teachers of African-American children*. San Francisco, CA: Josey-Bass.

Ladson-Billings, G. (1995). Toward a theory of culturally relevant pedagogy. *American Educational Research Journal, 32*(3), 465–491.

Lee, O., Quinn, H., & Valdés, G. (2013). Science and language for English language learners in relation to next generation science standards and with implications for common core state standards for English language arts and mathematics. *Educational Researcher, 42*(4), 223–233.

Loftin, C., Davis, L. A., & Hartin, V. (2010). Classroom participation: A student perspective. *Teaching and Learning in Nursing, 5*, 119–124.
McCarthy, J., & Benally, J. (2003). Classroom management in a Navajo middle school. *Theory into Practice, 42*(4), 296–304.
Montgomery Public School. (2010). *A resource for equitable classroom practices*. Maryland: Office of human resources and development, Montgomery Public Schools.
National Education Association. (2008). *Promoting educators' cultural competence to better serve culturally diverse students*. Retrieved from http://www.nea.org/assets/docs/PB13_CulturalCompetence08.pdf
National Research Council. (2000). *How people learn: Brain, mind, experience, and schools* (Expanded ed.). Washington, DC: The National Academies Press.
National Research Council. (2012). *A framework for K-12 science education: Practices, crosscutting concepts, and core ideas*. Washington, DC: National Academies Press.
Noguera, P. A. (2003). *City schools and the American dream: Fulfilling the promise of public education*. New York, NY: Teachers College.
Patrick, H., Ryan, A. M., & Kaplan, A. (2007). Early adolescents' perceptions of the classroom social environment, motivational beliefs, and engagement. *Journal of Educational Psychology, 99*(1), 83–98.
Phillips, M., Crouse, J., & Ralph, J. (1998). Does the Black–White test score gap widen after children enter school? In C. Jencks & M. Phillips (Eds.), *The Black–White test score gap* (pp. 229–272). Washington, DC: Brookings Institution Press.
Pringle, R. M., Brkich, K. M., Adams, T. L., West-Olatunii, C., & Archer-Banks, D. A. (2012). Factors influencing elementary teachers' positioning of African American girls as science and mathematics learners. *School Science and Mathematics, 112*(4), 217–229.
Reyes, M. R., Brackett, M. A., Rivers, S. E., White, M., & Salovey, P. (2012). Classroom emotional climate, student engagement, and academic achievement. *Journal of Educational Psychology, 104*, 700–712.
Richards, H. V., Brown, A. F., & Forde, T. B. (2007). Addressing diversity in schools: Culturally responsive pedagogy. *Teaching Exceptional Children, 39*, 64–68.
Spearman, J., & Watt, H. M. (2013). Perception shapes experience: The influence of actual and perceived classroom environment dimensions on girls' motivations for science. *Learning Environments Research, 16*(2), 217–238.
Swalwell, K. (2011, December 21). *Why our students need "equity literacy" [blog post]*. Retrieved from Teaching Tolerance at www.tolerance.org/blog/why-our-students-need-equity-literacy
Traynor, P. L. (2003). Factors contributing to teacher choice of classroom order strategies. *Education-Indianapolis Then Chula Vista, 123*(3), 586–599.
Wallace, T., & Brand, B. R. (2012). Using critical race theory to analyze science teachers culturally responsive practices. *Cultural Studies of Science Education, 7*(2), 341–374.
Weinstein, C. S., & Novodvorsky, I. (2015). *Middle and secondary classroom management: Lessons from research and practice* (5th ed.). New York, NY: McGraw-Hill.
Weinstein, C. S., Tomlinson-Clarke, S., & Curran, M. (2004). Toward a conception of culturally responsive classroom management. *Journal of Teacher Education, 55*(1), 25–38.
Wyatt, S. (2009). The brotherhood: Empowering adolescent African-American males toward excellence. *Professional School Counseling, 12*(6), 463–470.

Chapter 7
Metacognition: It's Thinking Time in Science

7.1 Thinking About Learning and the Sense Making Process in Science

Allison

I start each school year being hopeful for my students' science learning. This year, I am teaching a curriculum that was highly recommended in terms of its support for science learning and reformed teaching. I want to reach every child and often wonder about how and when to remediate and differentiate my science lessons to best support my class of low performing students. In my reflections, I constantly raise the following question: What do we need to do as science teachers to meet the learning needs of all students? Not everyone will study science as a career but I seek to lay the foundation for the development of lifelong skills and scientific literacy for all my students.

7.1.1 Introduction

I teach seventh grade science at a middle school in the southern area of the state. The diverse student population is composed of 34% white, 31% black, 27% Hispanic, 3% Asian, and 5% self-identified as being of mixed race (The Gold Report, n.d.). In addition, 2% of the total student population is comprised of English Language Learners (ELL) and 14% of the student receives Exceptional Student Education (ESE) services (The Gold Report, n.d.). Before the start of each school year and during summer planning period, a team of subject teachers and administrators review the student performance on the past year's state standardized tests. The results of the review are used to identify students requiring academic enrichment and guide their placements into appropriate classes.

In the last school year, 85% of the students scored below proficiency on the state's Reading and Mathematics assessments. While the school was labeled low-performing, this was the lowest achievement over the last five years. For the new year, one of the changes suggested by the review team was the organization of the

R. M. Pringle, *Researching Practitioner Inquiry as Professional Development*,
https://doi.org/10.1007/978-3-030-59550-0_7

group of 7th grade students by achievement levels for mathematics and science instruction. With the homogenous classes, teachers would be better able to provide individual and additional support per the needs of the students. Another advantage identified by the review team was that students in the higher performing classes would keep pace with the teaching guide provided by the district and be prepared for the quarterly assessments administered by the district. As the most senior science teacher and enrolled in the professional development program, I opted for the lowest achieving group of students as one of my assignments for the new school year.

My class with the lowest performing students consisted of eight boys and five girls. While this small class size seemed ideal for teaching science, the students had other challenges beside low achievement. These issues include eligibility for special services such as English Language Learners (ELL), students with disabilities (ESE), 504 plans to provide accommodations for ADHD, and engagement in the Response to Intervention (RTI) process, for behavioral and/or written language needs.

7.1.2 A Call to Action

The courses in our PD program were tailored to support our teaching of the reform-based science curriculum – Investigating and Questioning our World through Science and Technology (IQWST). This inquiry science curriculum was carefully developed to provide students with the opportunity to "investigate questions relevant to their lives by conducting investigations; collecting and analyzing data; developing and using models to explain phenomena, and engaging in argument from evidence, all in a literacy and discourse-rich environment" (Krajcik, Reiser, Sutherland, & Fortus, 2004). As I began to plan for my students, I specifically reflected on two courses, *inquiry-based science teaching* and *teaching* science to diverse learners. These courses were steeped in best principles and practices for science teaching with importance placed on surfacing students' past or lived experiences as a foundation for initiating conceptual change. This idea was also emphasized in the inquiry-based science course as we explored contemporary beliefs about how science is learned. Even with the knowledge of the students' low achievements, I had great anticipation about teaching science to the small class. I was especially looking forward to being able to surface their prior ideas about the science knowledge and to track the development of the new learning.

The curriculum provided strategies for teaching science and these were reinforced during the formal coursework in the PD. With the small class size, I had much interaction with the students and were able to institute the strategies in the curriculum. It was therefore with much expectation of progress when I assigned the first graded class assessment task. I was really looking towards even a minor glimpse that learning was occurring – some improvements in students' achievement. The students' performance was disappointing. I was concerned about their science learning and my hope about improving their learning at this point appeared daunting. It was clear the students were not understanding the science content knowledge and were not developing the skills as contained in the curriculum. There was an acknowledged issue in my science teaching and being able to facilitate students' learning.

As I began to organize for the second quarter of the school year, I reflected on the possible areas of the instructional sequence that needed to be adjusted to better meet the needs of my students. I tried to adhere as much as possible to the elements of effective instruction that undergirded the instructional sequence. The elements include motivation, eliciting students' prior knowledge, intellectual engagement with science phenomena, use of evidence to critique claims, and sense-making. As I reflected on the students' performance during the lessons, I realized that I may not have afforded them the kind of challenges that were expected in the sense making component of the lessons. Banilower, Cohen, Pasley, and Weiss (2010) state that students should be challenged to make sense of the content explored through appropriate engagement with the science phenomena being developed in the lesson. From my observation and upon reflection, I note that the sense-making aspect of the lessons were rarely conducted. The lessons usually ended without students being required to think, reflect, and make the necessary connections to the goals of the lesson. I wondered about my students and the role of sense making in supporting their learning. I also recognized that I needed to be intentional in developing this area of my lessons. This became the focus of my first teacher inquiry project.

7.1.3 Sense Making in Science: Developing Metacognitive Processes

In my teacher inquiry I set out to improve on the sense making aspect of my inquiry-based science lessons toward improving the science achievement of my class of low performing students. As I set out to conduct my teacher inquiry, I explored the literature to become familiar with existing teaching strategies. According to Dana and Yendol-Hoppey (2009), the literature is essential to teacher inquiry as it connects to, and provides information about the strategies being examined.

In the literature, there is a wide consensus that inquiry-based science teaching is effective in supporting meaningful science learning. During inquiry-based lessons, students develop and use scientific practices and learn scientific content (Furtak, Seidel, Iverson, & Briggs, 2012; Gibson & Chase, 2002; Keys & Bryan, 2001,). A report by Gibson and Chase (2002) concluded that inquiry-based science activities had positive effects on middle school students' science achievement, attitude and scientific skills when compared to traditional science teaching. Keys and Bryan (2001) further state that the effectiveness of inquiry-based lessons can be enhanced when teachers formulate patterns of teaching actions that are grade appropriate, and relevant to the science curriculum. With a focus on the instructional sequences during inquiry-based lessons, Banilower et al. (2010) explain that students "must be motivated to learn and intellectually engaged in activities and/or discussions focusing on what they already know" (p.5). They further contend that in science lessons students, should be engaged in intellectual work and be allowed to make sense of the new ideas through skillful questioning, class discussions, and opportunities to apply their learning to new contexts.

During sensemaking, students reflect on their prior ideas and become aware of how their thinking may have changed during the lesson. To achieve such, students will need to develop the skills to perform such activities. Psychologists and educators have identified metacognition as the set of skills that enable learners to become aware of how they learn. Many science educators believe that students reflecting on their learning is an important mechanism for promoting science learning. In a study examining metacognitive development, White and Frederiksen (2005) argue that developing metacognitive skills is crucial in inquiry-based science learning. They lament however, that it is unfortunate that little emphasis is placed on metacognitive skills in the nation's science curricular standards.

The review of the literature on metacognitive skills highlighted the inclusion of various forms of self-assessment and reflection, and providing students with explicit models of self-regulatory processes like planning and monitoring. In the IQWST curriculum, the framework already existed for students to experience science phenomenon, talk about it, reason and provide an explanation and be able to think about their own learning. In my teaching, my students were not explicitly being taught the metacognitive skills to support their learning during the sensemaking component aspect of the lesson. I therefore needed to make the necessary adjustments in my teaching practices to encourage the development of metacognitive skills among my students who were low achievers in science.

7.1.4 Teacher Inquiry

After observing the continued low science achievements and the ineffectiveness in my teaching, I became intentional in developing metacognitive skills within the context of the sensemaking process of the lessons. The literature provided some strategies that I proceeded to incorporate into the sense making aspect of my lessons. These included a self-assessment rating scale to gauge their learning against the learning goals, intense teacher questioning, and self-questioning, and reflective practices to think about what they were learning. As I began my inquiry project, I collected baseline information (e.g., FCAT scores, IEPs, and 504 plans) about the students. I kept a journal in which, I documented specific events, described the context, and reflected on each day's lesson. In addition, selected lessons were audio recorded and I collected artifacts of students' work. Students' work included responses on formative and summative assessments and digital pictures of consensus models developed during sense-making. At the end of the teacher inquiry period, students answered questions on a survey that provided feedback about their perceptions of the strategies, their learning progress in class, things they needed to work on, and the kinds of questions they were most likely to answer orally as well as in written form. The students' reflection provided insights into the sense making process and the metacognitive strategies used in the lessons.

Data analysis was continuous throughout the duration of the teacher inquiry. Each notable occurrence that emerged in response to my teaching, was used to inform the ongoing adjustments made to my practices. The process of coding allowed me to identify trends and patterns revealing pictures of the students' engagement during the sense making process within each activity. In addition, the quizzes and the final unit assessment allowed me to determine how well students were communicating their science content knowledge in writing, oral explanation, or through model drawing and interpretation. The assessment tasks included both directed prompts and open responses that provided another context for students to reflect on their learning and to demonstrate their knowledge. Their responses allowed me to conduct a systematic review of their development and to determine the extent of meaningful learning while also recognizing any gaps that was still evident in their learning.

7.1.5 Findings and Discussion

My teacher inquiry focused on the development of metacognitive skills during the sense making aspect of my science lesson. My goal was to study and adjust my teaching to better facilitate the science learning of my group of low performing students. The literature suggests that meaningful science learning will occur when students are able to think about and reflect on their own learning. I therefore incorporated sense making strategies into my lessons and in a systematic manner, observed and reflected on the students' responses as well as evaluated their learning. Data sources included my daily description of students engaging in the science activities and their performances on formative assessments, quizzes, and tests.

My teacher inquiry provided some interesting results. While the students enjoyed doing the science activities, they were not ready to be engaged in other areas of the planned instruction. For example, they had difficulties extending the hands-on activities into developing claims, evidence, and reasoning that was a natural progression of the lesson. They were not connecting the science activities, science content, and the learning goals, and struggled when asked to think aloud or to share their thinking. There was very little evidence that learning or even thinking about what they should be learning was occurring. I learned that for students who are low achievers in science, I first need to focus on the development of some basic skills before introducing them to metacognitive strategies and requiring them to think about their learning. Students need to be taught basic skills of observing, generating claims and being able to make justification based on their evidence.

Even with intense coaching, developing metacognitive strategies is difficult among low performing students already struggling with skills such as writing observation, and developing and justifying claims: At the start of the teacher inquiry, I allowed the students to share their thinking both aloud in class discussions or in individual writing and self-assessment using teacher provided prompts and rating scales. These strategies were gradually introduced and I offered support to individual students. Questions were also strategically placed in the lessons to initiate students' reflections on what they were learning. The questions were posed during both small and large group activities. For example, during a small group activity, I posed the following questions: What are you learning about chemical reactions? What questions do you have about what you are learning about chemical reactions? At first, I required them to write their answers on their worksheet and then to share with their team members. During the whole group discussions, some students were unable to share in a logical way what they did during the activities.

Sometimes amidst the excitement of doing the activities and reacting to their observations, students showed signs of understanding the science ideas. Even during these rare occurrences, the students struggled to reflect with accuracy on their thinking and hence their learning. This was more evident during sense-making when after completing the inquiry activities and whole group discussion, they were required to work in small groups and generate ways of representing their learning. The data analysis showed that the students did not respond to questions that required explanation based on what they did in the science activities and about their understanding of the science content that we were studying. As I reflected on my teaching and the level of success, I concluded that I was doing all I learned about an effective lesson, I covered all aspects of what would make a successful science class; but what do you do when the students do not have the requisite skills to engage in sense making? At first, I thought what was happening was due to a lack of motivation. Some researchers have identified a link between metacognition and motivations (Eiseneberg, 2010; Martinez, 2006). However, while the students showed excitement in mixing substances and observing the visible changes while doing the science activity, they were not able to monitor their thinking and or display other metacognitive skills.

Doing the science activities is fun but how do I move from the fun in the science activities to the independent learning: My students struggled with moving beyond the science activities in the lessons. A few weeks into the year, when the students entered the classroom, they automatically looked for signs that would indicate doing science activities. I frequently had to respond to the question: Are we doing science today? They enjoyed the mixing, pouring, and rubbing things together. With scaffolding during discussion and teacher notes on the overhead projector, they would write the observation on their activity sheet. I knew that the observations had to be in place if the lesson was to develop in a manner to get to the sense making. The students in most cases followed the instructions to "do" even though I had to monitor their how they were proceeding. Early in the project, I realized that because of the nature of my students, I had to develop the lessons in sections. Usually after the "doing" we would then talk about their observations. I would begin with, "tell me what you saw or heard?" Then, I would urge them to write down what they just said or heard. To offer support, I would project their responses. The class was small and I wanted the involvement of all the students. Each aspect of the lesson took a much longer time than planned because I wanted to maintain the spirit of having the students engaged in the process of inquiry – to get them to the point where they could independently respond to the questions guiding the lessons.

Having the students move beyond documenting the observation was a challenge. I had to use strategies such as scaffolding, applauding, and celebrating achievements to get the students to write and share their observation. For example, during the lesson on chemical reaction, I instructed the class to first write their own observations. I wanted the independent work before we had the discussion in their small groups or even in the whole group. The intent was that in working with their lab partners, they would have a more complete set of observation. This would be followed by a class discussion in which a representative would share the group's response. The lesson did not develop as planned. I had to use strategic questioning to elicit the observations and to project them while allowing some time for students to write in their activity books. After I had elicited the observation, one student volunteered to read aloud what she had written in her workbook – this was commendable and I had the class applaud her effort. At the end of the lesson, even though we had further discussion and the class was prompted toward making a claim, her workbook only had the observation she had read aloud.

In the same lesson, when students were asked to make a claim about chemical reactions, their responses became just reading the observations that we had discussed and they had written in their notebook. After much prompting, one student mentioned that "new substances are produced." He explained that the ball of aluminum foil changed into a new substance when placed in the beaker of cupric chloride solution and offered the change in temperature as evidence to support his claim. However, when asked to complete the writing task to express their learning, the same student's conclusion section in his workbook even with a direct prompt remained blank. Another student, who showed some understanding during the lesson also did not complete the response to the questions during the sense-making. The response of the class on a whole left me with more questions than answers – are these students ready to think independently as required in learning science?

7.1.6 Conflicts and Implications

The teacher inquiry process for me was an invaluable structured process, a "systematic study of [my personal] practice" in which teachers work within a framework for delving into the implementation of strategies to address student needs (Dana & Yendol-Hoppey, 2009, p.4). This systematic approach to teaching was personal and relevant to my teaching situation. As I researched the literature I was exposed to new strategies which expanded my pedagogical content knowledge. As I implemented these strategies with my students and analyzed the data collected during the inquiry process, there were valuable relevancy from the immediate feedback that served to inform my teaching. Dana and Yendol- Hoppey (2009) explain that "by participating in teacher inquiry, the teacher develops a sense of ownership in the knowledge constructed, and this sense of ownership heavily contributes to the possibilities for real change to take place in the classroom" (p.7). Within my own inquiry, I relied on my field notes, and reflections that I documented and analyzed around the strategies. During this time, I had students answer various kinds of questions and assessed their progress for sense making in various ways. This was a much more structured approach to determining the effectiveness of the strategies used for sense making than I had utilized previously in my many years of teaching.

I introduced strategies in my science lessons to strengthen the learning of low achieving learners. I attempted to get the students engaged in the processes in inquiry-based science learning. They struggled to make connections to what they already knew and the ideas being developed in the lessons. While in some cases they were able to articulate their observations, they experienced difficulties in evaluating and making claims based on the evidence that emerged during the science activities. I utilized various scaffolding strategies as I modeled practices and communicated expectations. During the sense making process, I varied the kinds of scenarios and ensured they were all relevant to the students' lived experiences. When I recognized that the strategies I was introducing were ineffective, I resorted to the extensive use of open-ended questions with the hope to include opportunities for critical thinking. For example, I utilized "Think-Pair- Share" a strategy that gives students time to think through their own ideas before having to present information to the class in discussion or in writing. The students also accessed mini dry erase boards to practice and edit their writing during discussions. This was then transferred to their notebooks before moving further into the lesson.

In my journal entries I noted that I was not being successful. At times, I began to wonder about my questioning techniques and the extent to which they needed some revising as "[t]he teacher's effectiveness in asking questions, providing explanations, and otherwise helping to push student thinking forward as the lesson unfolds often determines students' opportunity to learn" (Banilower et al., 2010, p. 13–14). However, the students' low achievement in science was more that the development of metacognitive skills. I returned to the literature and but there were conflicts with my observations. For example, Davis (2003) suggested that students reflect unproductively more often in response to directed prompts. I definitely did not find this to be true within my own teaching scenario. Also, according to the literature, the development of metacognitive skills begins in the elementary grades. My grade seven students were not able to engage in basic practices such as setting goals, thinking about their learning, or providing a reason for making certain choices.

7.1.7 Teacher Inquiry as Professional Development

As I reflect on my learning during the teacher inquiry, I have to keep at the fore my initial wondering of how to better assist students learning by introducing metacognitive skills. The occurrences during my teacher inquiry process challenged my perceptions of my teaching abilities. As a science teacher, I was confident in my abilities. I worked hard, I kept current with new strategies to further student success in my class, and I have experienced some success even with struggling learners. This teacher inquiry has confirmed that I still have a lot to learn about my teaching toward increasing science achievement. It is clear from my teacher inquiry that science learning among low achievers is not attributable to one factor. I thought that adjusting the sense making aspect of the lesson and focus on developing metacognitive skills would make a difference in science learning. From the intentional focus and adjustments to my teaching and immediate student learning, I learned that my students were not only low achievers in science they also had low language and literacy skills. I will therefore need to secure strategies for developing their skills and confidence as readers and writers to support their science learning. I have learned that translating strategies to address my problems of practice takes time and the results from my experiences may conflict with the literature. Upon reflection, the duration of the teacher inquiry may have been a factor in my level of success but I am now more apt to observe, reflect, and respond in the moment of my teaching.

7.2 Teacher Questioning and Student Thinking

Throughout my eleven years, I have participated in numerous district programs to improve my teaching. My immersion in this PD program is a continuation of my desire to be a more effective teacher. Learning from the courses have further heightened my desire to improve and experiment with teaching strategies and teaching techniques that could improve my teaching. Since I began this level of introspection and self-awareness, my questions and observations have led me to more questions and even more wondering about being an effective middle school science teacher.

7.2.1 Introduction

I teach middle grade science at a public school located in the western area of this southern state. The School currently serves approximately 1200 students in sixth to eighth grade and with approximately 35% Caucasians, 30% African Americans, 28% Hispanics, and 7% identified as others. The school received an "A" ranking from the State of Florida based on student success measures such as standardized testing and student achievement for twelve consecutive years. The school recently lost its "A" ranking. This lowering to a grade of "B" was the impetus for the administration urging of a much greater focus on learning for our students and in all subjects. The goals of the science program determined at the start of the school year was therefore to increase students' understanding of scientific concepts and ultimately their achievement. To achieve the goal, two science teachers were enrolled in the PD program and the school adopted the science curriculum which was an important aspect of the PD program. In the program, we were supported in our learning about the curriculum and in teaching our students. The curriculum was standard-based and carefully sequenced across the three middle school grades. Notable, the science ideas and science and engineering practices were coordinated across units within each grade level and across the three grades, sixth, seventh and eighth. This year my teaching assignment included one sixth grade class with plans to proceed as their science teacher across the three grades.

7.2.2 From the PD Program to Classroom Wondering

During one of our early face-to-face meetings in the PD program, I observed the professor's use of questioning. At first, my colleagues and I were disappointed when our questions became ours to answer and during the interactions we had to think about our learning and generate new questions. How can I think about my learning? How can I self-correct and monitor my own lack of understanding? These were questions I raised in my journal to deal which showed my disgust with the interactions with the professor. After these early interactions in the program, I began to realize the impact of the questions in evoking self-reflection and critical thinking about my own learning. Now, as I teach my students, I observe their struggles and their negative responses to questions that evoke reflection and critical thinking. They too did not welcome the times when they had to answer my questions or raise questions about their learning. In most cases, even the more advanced students wanted to be asked questions that did not require them to think. As I reflect on my observation, this was where I started in the PD program. However, as I learned more about inquiry-based teaching and the elements of effective instruction, and in many discussions with my colleagues in the program, I began to wonder about my use of questions to support the learning of my students. I recognized that the problem of my practice was my inability to use questioning as a tool to enhance science learning and the development of independent learners.

Sense-making is an important element in the development of science lessons. It is here teachers use skillful questions or critical tasks that allow students to make sense of the science experiences and connect to the learning goals (Banilower et al., 2010). During the courses in the PD program, I learned there was an art to framing questions in ways that would lead students to reflect on their learning. In addition, I need to be ready at all times to create more questions about the topic in an effort to improve their learning. For my science lessons, this meant the development of metacognitive skills. While metacognition includes questioning and thinking about one's learning, I wanted to develop my students as independent thinkers, observe my teaching, and improve my teaching strategies to better support their learning. My teacher inquiry was therefore about my learning as well as my students' science achievement. I set out to respond to the following question: In what ways can I use questioning strategy to support the sense-making process as students develop metacognitive skills during inquiry-based science teaching and learning?

7.2.3 My Cycle of Teacher Inquiry

Teacher inquiry began with the identification of a problem of my practice and the impact on students' learning. My focus was to examine the use of questioning as a strategy to become a more effective science teacher and increase my students' science achievement. My primary data source was audio recording of my lessons. I recorded five of my lessons using an app on my cell phone. In addition, I kept a classroom "diary" in which I wrote my thoughts after each lesson during the nine-week period. It became a forum into which I vented my frustrations, noted successes, and described the classroom discussions. The students also responded to an informal survey that elicited their overall perspectives about the classroom discourse and the nature of the intense questioning that were occurring. Their identifications were kept anonymously and were not linked to their responses. Finally, I interviewed five students in a focus group to get their perspectives about the curriculum and the use of questions during the lessons.

During data analysis, I listened to the audio recordings and identified key instances during my interactions with the students. From the audio recordings I was also able to gain some insight into how 'open' and 'closed' questioning affected the classroom discourse. Discourse analysis (Thwaite & McKay, 2013) became a valuable tool for analyzing the interview transcripts, reflections, and student artifacts. In the analysis, I also systematically examined the strategies, my use of questioning, and the responses of the students across all the data sources. The surveys also provided insights into the students' perspective regarding how questions were handled in the lesson. Finally, students' growth and academic gains were gleaned from the class assessment, student artifacts, and the interview.

7.2.4 Taking a Stance on Questioning in Science Classrooms

The teacher inquiry was conducted over a nine-week period as I taught Learning Set 3: How Does Moving Water Affect the Land? Data analysis showed that intense use of teacher questioning was a natural part of the inquiry lessons from motivating and surfacing ideas of the students through to their walking through the door at the end of the period. My attention to the importance of questions emerged from my own experiences as a learner in the PD program. The nature of the questions in the program and my interaction with the professor revealed the power of carefully designed and strategically placed questions in instruction.

At the start of the teacher inquiry, the sense making component of my instruction was occurring mainly at the end of the lesson. The lesson however was usually interrupted by the bell that signaled the end of the teaching period. Overtime, as I began to focus on questioning as a strategy to support learning, sense-making became a natural component of all areas of the lesson's development. I gleaned a great deal of information about science teaching as I made the adjustments to use questions to develop students' thinking throughout the lessons. The increase in sense-making activities required carefully, pre-planned and well-crafted questions but also much flexibility in my teaching. The pace of the lessons was also impacted and this had implications for completing the sixth grade curriculum.

7.2.5 Can I Break Free from the Planned Curriculum?

I began the teacher inquiry with the intention of using all the pre-determined questions in the curriculum. At first glance, they seemed to be different types of questions and their introduction into the lesson usually followed a pattern. The first three questions asked students to recall information from the previous lesson and the later ones required the students to go beyond their observation and to engage in reasoning and providing explanation. Once I started to focus on my observations, I realized that when I kept to the pre-planned questions my lesson appeared robotic. There was a slow motion to the development of the lesson as students were mostly unresponsive to the questions and "wait-time" did not achieve an increase in their responsiveness. I soon abandoned my total dependence on the suggested prompts. This was evident in the audio recordings where in one lesson, I counted 57 teacher questions. My questions at first followed the strategies identified the curriculum – directly related to getting students to share what they were doing in the lesson and to share what they were thinking about their own learning. As I reflected on the interactions, I realized that some of the questions required obvious answers. Obvious, because if the students were following the lesson's activities, whether observing or doing it themselves, the answers would be right in front of them. In my journal, what was obvious was my questions were guaranteed for class management. "Were the students doing the activities as they should? If they were, then they easily had the answers," I wrote in my journal. For example, in the lesson on how water affected the earth, as students were observing the water flowing down the stream table, I asked five questions and they were all about what they saw: What is happening as the water flows? Would you say the water is moving fast or slow? Where is the water going? As I listened to the questions, the management and control was obvious as the questions were usually followed by chastising students who were drifting away from the area or engaging in other behaviors not related to science learning. I however consoled myself that I was ensuring they were making observation – collecting evidence. In the total interaction, there was no space for students' questions about the phenomenon they were observing and also missing were the kinds of questions to forge the development of critical thinking and metacognitive skills. Acting on my observations, my questions began to require students to share what they were thinking and to identify questions that were still unanswered. I therefore resorted to small group discussions and their collective response was to develop one question to ask another group.

One of the learning goals of the lesson was for students to generate questions about the processes that affect materials on earth. In one of the lessons, I missed an opportunity to provide the students with on the spot sense making experiences. As I listened to the audio, I wrote in my journal about the missed opportunity, "this is Florida, near the coast, and last year a hurricane passed through." Upon reflection, I realized I missed the opportunity because the sense-making process is the last of the elements of effective instruction and even though I was making changes, I was locked into the planned curriculum. My ability to respond and be effective in the use of questions as a metacognitive strategy became limited by the instructional plan of the curriculum. By the fourth week of data collection I began to get more creative with my methods of questioning. For example, during formative assessments, I introduced specific questions that required students to provide written responses. In addition, they had to provide one question for display on the driving question board. I didn't consider this move a major alteration to my typical assessment procedure but it added another opportunity to develop thinking skills among my students.

7.2.6 *It's Not a Test: Getting Students to Think About Their Own Learning*

The purpose of the Driving Board as a feature of this curriculum was to display the essential questions of the unit and also provide a space for students to raise their questions. Without prompting, students failed to generate questions from the science lessons. In one of the interviews, students indicated that if they had questions that did not get answered they usually did not write them down. They said that if they had questions, the best way to get answers was to ask the teacher. This was the thinking among my students. The reason that this data point is significant is that I am trying to develop metacognitive skills in the students that would promote their thinking about their learning and surfacing question for further inquiry. This inquiry could be on their own or opened up in ways to include the class by placing them on the Driving Board. To achieve this, a certain level of academic independence must be reached and this required me to be strategic in engaging the students. I then set out to create a classroom in which the discourse was being driven by students. Achieving this level of discourse was important as I sought to be more effective at using questions to develop reflective and thinking skills in the students. When students ask questions, they are shaping and exposing their thoughts (Watts, Alsop, Gould, & Walsh, 1997) allowing me to further scaffold their learning. The process of shaping and exposing their thinking would support the students' development of deep conceptual understanding of the science they were learning. In addition, I wanted my questions to not only facilitate students' conceptual understanding, but was hoping to guide them toward valuing their own thinking and the sense-making that was occurring. I wanted to develop independent science learners.

I continued to observe the students' response to oral questions. Upon reflection, I realized asking oral questions to engage students in sense-making and leading to worthwhile discussions was ineffective in initiating self-reflection. Oral questions also did not provide enough support and scaffold for the students to identity what they know and what they want to learn after they had gone through the lesson. During the last segment of one of the lessons, I decided to add another layer to the sense-making process. I generated some questions as tasks that would require reflection, written responses, and connections to the learning goals. I titled the paper sense-making and as I proceeded to pass them out to the class, I told them I would entertain their questions before they began the tasks. "You did not tell us we were getting a quiz. How many questions are we getting" one of the boys asked me seconds before I handed the paper to the first student or even gave the instructions. I projected the learning goal which was to: Apply processes of weathering and erosion to change over time. Students were then asked to examine the learning goal as I challenged them to think about the activity that was conducted in the lesson and to reflect on the ways they thought the activity was connected to the learning goal. As usual, hands were quickly raised and students asked questions that were mainly mechanical and not related to science learning – one of which was, is this really a test? To calm the students, I started to negotiate with them and gave them the opportunity to work in pairs to discuss the relationship between the lesson's activity and the learning goals. That however was my real intention – to have them work in pairs, reflect on the lesson, and together think about their learning. Interestingly, as they worked, I walked around the room and began to glance at what they were writing. As I approached one pair of students, I realized they were intently watching my face as I read their statements. I suspected they were trying to figure out my thoughts or for me to "tell" them they were on the right track. Honestly, I couldn't help but grin at their cunning efforts. I am sure the response from my body language gave them the "answer" and as I slowly moved away and onto the next pair without making a comment. In an informal interview after the class, the students said that this was the first time I looked at their work and did not ask a probing question. This they said got them further thinking because they had to make the decision based on the conversation with their partner and not being influenced by my questions.

As I reviewed the students work, they had many scribblings and markings through their writing. In most cases, the markings seemed to represent their feelings as they moved from being angered and ignored to being challenged and amused by the time they sauntered out of the room. On one paper the scribble was, we do/not have a question. Below this statement was written: Do all tiny rocks come from large ones? What was interesting here was the number of lines running through the scribble. When I compiled and read the responses, I recorded their questions to be used to start the next science lesson. In my journal, I scribbled "for the first time in my class period, I had the ideal classroom where the students were working harder than the instructor. They were engaged in focusing and unpacking what they truly had learned." The pairing of students and providing specific tasks became a stable activity in my class. As I planned the tasks, I also had to pre-develop higher order questions that would not require easy answers but allow the students to collaborate with their peers in reflecting and thinking about their own learning. Ironically, getting to a level of comfort in developing these questions became easier as I learned how to rewrite questions suggested in the lesson plans. This for me was a level of teacher empowerment.

7.2.7 Conclusion and Implications

My teacher inquiry was about my teaching and learning and making the changes to be better at developing my students to become independent learners and being able to think about their own learning. I came to this realization early that every question raised about my teaching led to other questions. Each observation opened another door of observations and the improvements that can be made to hone my craft as a teacher became continuous. This systematic reflection on my practice through teacher inquiry has been truly eye opening for me. When I first set out with my wonderings, I felt that the whole notion of "questioning" was a topic that though difficult to pursue would make a difference to my teaching. The teacher inquiry was as much about me as well as my students learning. What I learned from my teacher inquiry is that questioning, if used in a strategic manner, can pique students' interest regardless of the scientific concepts all the while giving them ownership of the discourse and allowing them to think about their own learning. The IQWST curriculum suggested question prompts to be used during the course of a lesson. Typically, these questions are in the "synthesizing" category of knowledge requiring extended thinking. As I inquired into my use of questions, sometimes these planned curricular questions became barriers to engaging the students in ways for them to think about their learning. As I made the adjustments, the discourse in my classroom took on a more "organic" and natural flow. One of the most eye-opening learning from my inquiry is that strategic questioning can be an extremely useful tool in supporting metacognition and dictating classroom management. Teacher inquiry never ends but is a continuous cycle that defines a reflective teacher.

7.3 Learning How to Think: Metacognition in Science

Britni

This is my fifth-year teaching science at a public school in this county, but I have a much longer history with the district as a student, having attended public schools for 13 years. However, since I was a student a lot has changed: No Child Left Behind legislation has raised the stakes on standardized tests, Sal Khan made it possible for teachers to flip their classrooms, and smart phones allow information access in seconds from nearly anywhere. Today's students need to master a dizzying array of technical skills and knowledge and a thirst for continuous learning. That is, we need to teach them how to fish rather than giving them a fish. But, how do we achieve this in an era of standardized tests that focus on the fish?

7.3.1 Introduction

I am a sixth grade science teacher at a community middle school in a large school district in the state. Our school is highly diverse: 85% are minorities, over 70% are bilingual, and 25% are classified as English Language Learners (ELLs). The school embraces the importance of developing the whole child and in addition to academic subjects, has a wide range of activities that identify and allow the students to develop skills such as photography, public speaking, drama, and more to be ready for high school. My school has been deemed low performing for many years based on the state's evaluation. As I teach science in the sixth grade, I am expected to lay the foundation for science learning and students' performance through to the eighth grade where the state test in science is administered. The hope for this year and going forward is that students' achievement in science will increase.

I was selected to participate in this PD program by my school administrator. My involvement in the program was to improve our science teaching and learning at the sixth grade. My assignment included teaching a reform-based curriculum – Investigating and Questioning our World through Science and Technology (IQWST) that would be supported by activities in the PD program. IQWST is designed around the principles of coherence and progression so that students can, over a school year and across the three years of middle school develop a deeper understanding of the nature of science (Shwartz, Weizman, Fortus, Krajcik, & Reiser, 2008). This reform curriculum allows student to do and learn science by inquiry and in the process develop and use science and engineering practices. These practices in the process of science learning allow students to use prior knowledge as they synthesize and accommodate new ideas (Roseman, Linn, & Koppal, 2008).

7.3.2 Wondering

My school has a history of low performance on the state's science test administered at the eighth grade. Over the years, I have also observed the sixth grade students' struggle with science learning and this is seen in their low science achievement scores on classroom and district's test. During our science department meetings, a constant observation by the teachers is that the students enjoy doing science activities but are mostly unable to demonstrate an understanding of the science ideas. As a team, we have worked in collaboration and reviewed students' class work, including formative assessment to identify and address areas of weaknesses. Regardless of the grade level, our students have consistently demonstrated a lack of some skills to support their science learning. For example, they do not know how to analyze the information, how to ask questions, synthesize their own knowledge, how to think critically, and use revisional thinking. Some of these skills and abilities related to the concept of metacognition are specifically developed in the lessons in IQWST and are discussed in our courses in the PD program. As I made sense of metacognition in my own learning, I recognized that students were not able to "think about their own thinking," engage in self-reflection, and were not poised or empowered to become independent learners.

As I approached teaching IQWST, I wondered about student science learning and the idea of focusing on the development of their metacognitive skills. Many of the students were usually excited about doing the science activities but were unable to demonstrate meaningful learning of the science ideas. If I develop the metacognitive skills of the 6th graders, will they become better science learners? The literature suggests that sixth-graders should be developmentally capable of developing metacognition but I was not sure if I was capable of integrating metacognitive skills into IQWST or, if I could make it work in my classroom with all the other curricular requirements. My wondering is therefore about the extent to which I can integrate the teaching of metacognitive skills in ways that support students' learning. The questions guiding my teacher inquiry became: What strategies will develop student's metacognitive skills within the context of inquiry-based science? How will my teaching change as I include metacognitive strategies in my science teaching?

7.3.3 Wading Into Metacognition: My Cycle of Inquiry

My teacher inquiry was about examining how I integrated metacognitive strategies into inquiry-based science lessons. I began by exploring the research on metacognition in the literature. My goal was to develop a good understanding of metacognition and identify and select strategies to incorporate into my teaching. This was the start of my cycle of inquiry.

7.3.4 What Does the Literature Say About Metacognition?

Metacognition is important in science learning. Researchers have given much credence to the importance of metacognition for regulating and supporting students' science learning (Gunstone, 1994; White & Frederiksen, 2005). In simple term, metacognition is defined as "thinking about thinking" (Flavell, 1979). Some of the constituents of metacognition discussed by psychologists include self-monitoring of one's learning, goal setting, reflection, and being able to activate prior knowledge (Flavell, Miller, & Miller, 2002; Schraw, Crippen, & Hartley, 2006). Metacognition is complex but according to educators, metacognitive skills can be taught as long as students are afforded support and appropriate learning experiences. Research has shown that children can be taught strategies to improve their ability to reflect on their thinking, improve their understanding by engaging in self explanation, activate background knowledge, and plan ahead (National Academies of Sciences, Engineering, & Medicine, 2018; Tusuzoglu & Greene, 2014).

In science teaching and learning, metacognition plays a major role in achieving conceptual understanding and can support the application of new knowledge to new settings (Gunstone, 1994). With improved metacognition, students will be better able to learn inquiry-based science as they consciously engage in science and engineering practices, monitor their progress and develop an understanding of science ideas (Thomas, 2012; White & Frederiksen, 2005). According to Thomas (2012), classrooms oriented toward the development of metacognitive skills are identified by certain characteristics. These include metacognitive demands on students; discourse regarding learning and activities that enable successful learning; students having adequate levels of control and choice in relation to those activities, and emotional support and trust between the teacher and students. Other researchers indicate that teaching metacognitive skills are best achieved when instructions include diverse techniques within the context of the lesson being developed. Metacognitive strategies if developed accordingly can shape and impact learning students' learning (Metcalf & Kornell, 2005). A review of the literature clearly shows that if the goal of science teaching is to support conceptual understanding, it is important to teach students to reflect on their thinking, connect to their prior knowledge, and develop skills such as self-regulation.

7.3.5 Data Collection and Analysis

To respond to my teacher inquiry questions, my data sources included four summative assessments, field notes, student class work and note book, formative assessment probes that were given three times over the course of the unit and pre and post assessments for both content knowledge and metacognitive abilities. I adapted the Metacognitive Awareness Inventory (Jr. MAI) (Sperling, Howard, Miller, & Murphy, 2002) to collect information about two factors thought necessary to successfully utilize metacognitive strategies: knowledge of cognition and regulation of cognition. The Jr. MAI inventory, specifically gauges the metacognitive processes of middle-school-age children. As I respond to students' low science achievement, my teacher inquiry focused on my observation and the changes I made to my teaching. I also kept a journal in which I wrote my daily observations and reflections on my practice and the interactions with students. The journal entries also included a record of my own thinking during the entire data collection period. That is, I monitored my own metacognitive development and the ways I engaged the students in the strategies.

Data analysis occurred in different phases. I compared the students' responses on the pre and post assessments and examined the various artifacts. For example, a review of their models developed overtime, provided evidence of how they had revised their thinking. The students' note book allowed me to review their explanations and the formulation of their understanding. Finally, the students' responses to the series of assessment probes over the course of the unit revealed the extent to which individual students changed their thinking in light of new experiences.

7.3.6 *Findings and Discussion*

My teacher inquiry was conducted during my teaching of the chemistry unit. In this unit, students examine phenomena and build science connections around major questions that connect the big ideas Matter in chemistry. These lessons also require students to analyze data, and share their newly constructed knowledge about "the particulate nature of matter." As I teach this unit, I set out to use several strategies garnered from the literature to deliberately introduce students to, and develop their metacognitive skills. Strategies included surfacing prior ideas, revisional thinking, ongoing use of formative assessment probes, extended time to reflect on meaningful cognitive activities with opportunity for revision, and goal setting with graphic representations of progress. The strategies were intended to encourage students' learning as they tracked their own thinking and gauged their learning against the learning goals of the lessons. The IQWST curriculum contains many opportunities for students to observe phenomena, generate evidence-based claims, and think critically as they offer reasoning and explanation. One of the strategies I implemented during my teacher inquiry was "revisional thinking." As I introduced revisional thinking, students were encouraged to look at their previous ideas or past versions of their own work and reflect on the progress or changes to their learning.

At the start of the project, I observed students forgetting what they had identified as their previous knowledge sometimes at the start of the lesson. This became an issue during the sense making aspect of the lesson when I required them to compare their newly constructed learning with where they were at the start of the lesson or the unit. To overcome this issue, I required students to write their previous ideas on note paper secured in their workbook. Sometimes, I had them write the date and the time they were recalling the information. During the sense making or after the discussions that preceded an activity, I would provide the following sentence stem:

At first I thought ________, but now I know________.

This was usually done on a strip of paper that I provided and after completion, pasted into their notebook. Once my observation showed students having a level of comfort with the revisional thinking, I shifted the focus onto scaffolding model building.

The development of models was a key scientific practice developed in the sixth-grade curriculum. As students conducted their inquiry-based lessons including the examination of science phenomena, they constructed models to represent their observations and developing explanations. In the literature, developing models and refining with each additional idea, is considered a cognitive activity. As the lesson or the unit developed, the students were allowed to revisit and revise their models in response to new ideas. The revisions usually occur until the class agrees on a consensus model. The initial model reflects students' preexisting ideas and served as the starting point for conceptual change. For example, in a lesson investigating gases, students investigated the properties of air by observing the movement of the plunger in the barrel of a syringe. Students drew models of the air to first represent their observation. The next model was developed after the observation and served to explain what was happening to the air when the plunger was pushed and pulled in the barrel. Overtime, as more ideas were introduced, and as the class did more observations and discussions, students revised their models to better represent their understanding. Finally, as a class we arrived at a consensus model agreed on by the class. The consensus model represented the properties of gases including how and why they react to an increase in pressure and in relation to volume.

At each stage of the process, the students examined, reflected, and revised their models leading to a more valid representation of the phenomena being studied. It is this building and rebuilding of the model that was effective in deepening their conceptual understanding. During model building, the students needed additional time to process the information and make the necessary connections as their ideas changed in response to the lessons. In addition, as I adjusted my teaching, I required students to list the changes they made to their model and to write one convincing argument about why the changes were necessary at each stage. In addition to being strategic and in framing the metacognitive strategy around the already cognitive requirements, I had to constantly allow additional time for students to reflect and think about their own learning – this greatly impacted the duration of the lessons and keeping up pace with the guidelines from the district.

As I conducted the teacher inquiry, one connection I made between my own teaching and my classroom practice was the importance of reflection in every area of learning. It became even more important as I reflected on my teaching on a daily basis. Sometimes, immediately after each lesson I wrote my observations, recreated interactions in some places, and engaged in intense reflection. But my students needed to learn how to reflect. Reflection was important to expedite positive changes in student learning. I used a number of reflective strategies with the students. These include self-reporting assessment where students rated their own level of mastery on a 1–5 scale where five represented full understanding, and formative assessment probes (FAP) suggested by Keeley and Tugel (2009). When I first introduced and reviewed the students' ratings of themselves, I compared the individual ratings to their responses on other assessments. The scores were not aligned as students consistently gave themselves higher ratings. In response to the self-rating, I made one adjustment requiring them to provide their rating along with completing the following sentence stems in relation to the learning goals:

I rate myself a ________ because I understand __________.
I want to know more about _________________ before we do the test.

Introducing the completion of the sentence worked! The ratings showed some differences but gave a good picture of what the students were learning. Through practice with self-assessment ratings, students became increasingly more accurate at determining how well they could complete the targeted learning goal. Still, the lower achieving students continued to self-rate higher than the more advanced students whose performances were above average.

Formative assessment usually gives the teacher an indication of students' learning as the lesson or unit is being developed. In the teacher inquiry, I experimented with the use of formative assessment probes as a means to elicit prior knowledge and ideas, monitor changing conceptions, and give students an opportunity to reflect on how their own understanding changed over time. For example, in one of the probes, students were given a scenario that required knowledge of the particle nature of matter to correctly explain the phenomena. Students then selected one of six responses related to the phenomenon being investigated and wrote an explanation for their selection. Students were then given the option to represent their thinking about the phenomenon and drew pictures to support their explanations. The FAP was administered three times during the unit. The first, elicited their prior knowledge and as the unit developed it monitored students' changing conceptions and allowed them to reflect on how their own understanding changed over time.

In the sense making process of the lessons, students engaged in whole and small group discourses as they constructed and revised their science ideas. As I observed my students during the lessons and adjusted my teaching, I narrowed the number of strategies and focused on improving the surfacing of their prior knowledge and its use in the development of the lessons. I therefore provided cues such as: "I used to think … but now I know" to guide their reflection. I also varied the points in the lesson when strategies were interjected.

Before the teacher inquiry and consistent with the development in the curriculum, the sense making aspect of the lessons were usually toward the end of the teaching period. This resulted in the rushed closure of the lessons usually at the last three minutes of class with little or no opportunity for reflection. During this inquiry, I was able to much more readily recognize and emphasize the areas of the curriculum aimed at metacognitive development. The teacher inquiry with its systematic observation and reflection caused me to direct my focus away from getting through the big ideas only toward allowing my students to develop skills that would support their conceptual understanding and help them become independent learners. That is, I focused on students' learning above the need to be on pace recommended by the district.

In the final class assessment, my review showed there were some changes in students learning. The results for the FAP over time provided some interesting results. Some students' responses developed positively in a linear manner over the three times they worked on the assessment. Some students (8%) provided a valid explanation in the second administration and at the end they made changes that resulted in explanations that were not supported. Overall, the students who were higher performing rated their metacognitive processes higher than their other peers prior to the explicit instruction and implementation of metacognitive strategies. The analysis of the Jr. MAI self-report of metacognitive skills revealed that the lower performing students benefited most from the metacognitive strategies – overall, they identified learning gains attributed to the metacognitive strategies. Interestingly, it may be that lower performing students got more from the explicitly taught strategies than do higher performing students. The results were that metacognitive strategies can be implemented successfully with sixth-grade students enhancing their science learning.

7.3.7 Conclusion

Metacognitive strategies are supportive of science learning. The strategies help students learn how to track their thinking and gauge their learning against the learning goals and standards set for them. Some metacognitive strategies were already incorporated in the curriculum but were not explicitly defined or connected to the concept. These were easily overlooked as I developed my lessons as presented in the teacher's edition. Upon reflection and having learned from the teacher inquiry, the curriculum did not place that much emphasis on the develoment of metacognitive skills, or so it seemed. However, as I adjusted my teaching strategies in response to what I was learning, students benefited mostly from the metacognitve strategies that were framed around meaningful activities. Furthermore, emphasizing the ways they surfaced their prior knowledge and requiring them to reflect on and compare their new knowledge worked well. My teacher inquiry heightened my awareness that middle school students are capable of developiing metacognitive skills. These skills can be improved over time with consistent and specifically tailored instructions. However, metacognition may be difficult to implement into the classroom without mastery of time management or flexibility in extending the curriculum. Learning is more effective when students are allowed sufficient time and scaffolding toward reflecting and thinking about their learning. Students cannot engage in meaningful reflection when they haven't been given adequate time to process before moving on to the next lesson.

7.3.8 Teacher Inquiry as Professional Development

Teaching allows practitioners continuous opportunities to respond immediately to issues of practice in their classrooms based on their observations. The goal of the continuous adjustment is about improving students' learning. Being engaged in the teacher inquiry has given me a new perspective in dealing with the complex and challenging processes of being an effective science teacher. Through this process, I have researched many metacognitive strategies, implemented a few, and learned how they can be used to improve students' learning. Adjusting my teaching strategies to include the development and use of these metacognitive skills allowed me to view teaching through a new lens – the lens that includes observation, reflection, practice, within a continuous cycle leading to transformation. Simply put, teacher inquiry for me is a continuous process of my learning and making the adjustments to my teaching to make learning better for my students. Good teachers can become great teachers by looking through a myriad of lenses at their own practices and this is facilitated through teacher inquiry.

Most current PD program are plagued with issues about learning that are similar to learning in science classrooms. That is, learners have a hard time connecting the new information to their own experiences (or classrooms). The experiences may not be interesting or motivating, the topic may be completely unrelated to the interests and needs of the teacher, and there is no attempt at developing metacognitive strategies. Through the teacher inquiry, I became empowered to shape my own classrooms according to the students and their needs, not what was specifically mandated by state politicians or county officials. Overall, this was an empowering experience that will inform my practice going forward as I observe my students and wonder about the ways I can better meet their learning needs.

7.4 Researching Teachers' Inquiry: Metacognition, It's Thinking Time in Science

Developing metacognitive skills is an important concern for teaching and learning science. This is especially true in the era when the goal of K-12 science education includes the development of individuals with sufficient knowledge of science to participate in public discussions; are careful consumers of scientific and technological information related to their everyday lives; are able to continue to learn about science outside school; and have the intellectual skills of thinking and reasoning to enter careers of their choice, including (but not limited to) careers in science, engineering, and technology (Hodson, 2003; NRC, 2012). Regardless of the teaching context, there is an agreement that twenty-first century science learning experiences should include a coherent instructional sequence with rigorous goals (Krajcik, McNeill, & Reiser, 2008), and should support the integration of knowledge, dispositions, and science and engineering practices. Such science instruction must not only seek to achieve curricular goals but also develop metacognitive and lifelong learning skills needed to succeed at higher levels of science (Schraw et al., 2006).

Metacognition, the ability to monitor and regulate one's own cognitive processes is a complex construct that involves both knowledge and self-regulation (Kuhn & Dean, 2004; Schneider & Lockl, 2002). In general, highlighting the active nature of science learning, scholars agree that metacognitive processes underlie the learner's ability to monitor his or her own current level of learning and understanding (National Academies of Sciences, Engineering, and Medicine, 2018). Furthermore, described as a multidimensional set of skills that involve "thinking about thinking," at the very least, metacognition should be a high priority for science teaching. Thomas (2012) identifies metacognition as key to attending to the multiple agendas in science education. These agendas include the development of students' scientific literacy and their understanding of the nature of scientific inquiry, the nature of science itself and science concepts. However, the problem of learning to learn is complex and represents a potential challenge to science teachers.

The three teachers, Alyson, Anthony, and Britni were each teaching middle school classes that had students who were persistent underachievers in science. They observed their students' inabilities to activate their background knowledge, verbalize their thinking, and monitor their understanding toward the learning goals. After reflecting on their teaching practices, the intentional focus of integrating metacognition into their instruction was identified as a possible strategy to address the students' low achievement in science. Each expressed the goal to help students develop metacognitive skills that would foster conceptual change and increase their science achievement.

Approaches to science teaching and learning are now directed by substantive research from interdisciplinary communities including education, cognition and developmental psychology. In contributing to our current understanding, the researchers contend that the process of learning involves the adaptive interaction among existing knowledge or beliefs of the learner and new learning experiences from which knowledge will be generated. Further, learners are more likely to learn when they are able to reflect on their own thought processes and are aware of their thinking and learning (Duschl, 2008; Duschl & Grandy, 2008; Duschl & Hamilton, 2011). Furthermore, the researchers contend that students must be motivated to learn and intellectually engaged in activities and, or discussions which begins with what they already know. Contemporary learning theory suggests that students will best understand science content and the scientific process, if teachers encourage the development of certain practices. Students need to learn how to use evidence to support their claims and be helped to make sense of new, developmentally appropriate ideas in the context of their prior thinking and their understanding of related concepts.

Children have a natural curiosity about their world. In their everyday interactions they have amassed a set of ideas and conceptual framework that, from their experiences, explain how the world works. It is this natural curiosity and prior knowledge brought to the science classroom (NRC, 2007) that will play an important role in their science learning. Research suggests that learning, an active subjective process, is shaped and structured by the learners' past experiences and new knowledge is generated through intellectual discourse rather than passive reception of information. In this process, students as active processors of information, perform such crucial activities as observing, asking questions, generating evidence-based claims, rereading difficult material, and during reasoning provide viable explanations. The teachers were enacting an inquiry-based science curriculum that embraced the hallmarks of current reform efforts including an instructional sequence guided by elements of effective instruction. In discussing the elements of effective science instruction (Banilower et al., 2010), laud strategies that are metacognitively oriented and in agreement with Thomas (2012), they place much emphasis on science learning environments in which there is an appropriate level of metacognitive demands on students.

A plethora of documentation in the literature supports the development of metacognitive skills in the elementary school years (Flavel, Miller, & Miller, 2002; Harris, Santangelo, & Graham, 2010; Whitebread et al., 2009). The researchers agree that developing these skills in the lower grades will support dramatic forms of conceptual change in students as they transition into high schools. As with other elements of effective instruction, developing students' metacognitive skills may be accomplished with a variety of teaching strategies that are carefully implemented during instruction.

The importance of providing quality learning opportunities for increasing awareness of ones learning and the capacity to direct it, is echoed by a number of educators (Kramarski & Mevarech, 2003; National Academies of Sciences, Engineering, and Medicine (NASEM), 2018). As the teachers revealed in their narratives, science learning environments are a buzz of events as they enact instructional sequences. In the process, students monitor new ideas and compare with their previous knowledge and make connections between what they did in the lesson and what they were intended to learn. Students also connect the new experiences to what they have learned previously thus, placing the learning goals in a larger scientific framework as they organize the new knowledge in their cognitive framework (Gallagher, 2000; National Research Council, 2012. Finally, students may be given opportunities to apply the concepts to new contexts; this helps to reinforce their understanding of the ideas and build their reasoning and critical thinking skills. The teacher's effectiveness in asking questions, providing spaces for students to reason and offer explanations, and otherwise helping to push students thinking forward as the lesson unfolds often determines the quality of opportunities to support students' learning.

The agreement in the literature is that inquiry-based science teaching is effective in supporting science learning and has the potential to contribute significantly to the development of metacognitive skills. The teachers in adjusting their teaching, introduced students to metacognitive skills during the lesson and within the development of the subject matter (White & Frederiksen, 1998). They however adjusted their teaching in response to their observation. For example, teachers used multiple entry points and different approaches to introduce the metacognitive skills into the lessons. Yet, as described in the narratives, the middle school teachers encountered practical problems in achieving meaningful engagement of their low achieving students. While the students displayed levels of enthusiasm in "doing" the science activities, they had issues pertinent to the state of their learning. Students had low proficiency with literacy skills such as reading and writing, were unable to articulate or surface their prior knowledge, and could not construct themselves as science learners.

In inquiry-based science, students examine phenomena and when constructing scientific explanation, use data as evidence to support their claim. Their engagement in these science practices have the potential for them to develop reasoning, analytic, and critical thinking skills (Duran, 2016; Lawson, 2010). This was the essence of the IQWST curriculum that the teachers in the PD were teaching while at the same time were being supported in their own learning of how best to develop the lessons. The onus however was on the science teachers to translate the curriculum to support meaningful learning. As the curriculum is translated into practice students' reasoning and beliefs are made visible, and they are afforded relevant and appropriate levels of assistance to scaffold the development of metacognitive skills.

The duration of the teacher inquiry projects was approximately nine months. The teachers recognized the importance of adjusting their instruction as they seek to develop the metacognitive abilities during inquiry-based science teaching. Their inquiry projects highlighted the fact that introducing metacognitive skills to low achievers requires much more maneuverings than contained in the literature. The teachers' experiences provide a lens that amplifies real world occurrences in science classrooms. The teachers identified their problem of practice and set out to challenge themselves to make the necessary adjustments to their teaching, in ways to support students learning. In the end, their experiences included struggles and more concerns about increasing students' science achievement. The learning emerging from their experiences and captured in their narratives has implications for science teacher education. Notably, science teachers need to develop a repertoire of teaching skills and a toolbox of strategies that would allow for flexibility and offer more choices in meeting the needs of the learners. The curriculum in science teacher education should include teachers' development as metacognitive experts within the context of reform-based science teaching. Armed with this level of expertise, teachers will be better able to facilitate the development of such skills in our students.

References

Banilower, E., Cohen, K., Pasley, J., & Weiss, I. (2010). Effective science instruction: What does research tell us? *Educational Researcher, 33*(8), 3–15. Center on Instruction.

Dana, N. F., & Yendol-Hoppey, D. (2009). *Facilitator's guide: The reflective educator's guide to classroom research: Learning to teach and teaching to learn through practitioner inquiry*. Thousand Oaks, CA: Corwin Press.

Davis, E. (2003). Prompting middle school science students for productive reflection: Generic and directed prompts. *Journal of the Learning Sciences, 12*(1), 91–142.

Duran, M. (2016). The effect of the inquiry-based learning approach on student's critical-thinking skills. *Eurasia Journal of Mathematics, Science & Technology Education, 12*(12), 2887–2908.

Duschl, R. (2008). Science education in three-part harmony: Balancing conceptual, epistemic, and social learning goals. *Review of Research in Education, 32*(1), 268–291.

Duschl, R. A., & Grandy, R. E. (Eds.). (2008). *Teaching scientific inquiry: Recommendations for research and implementation*. Rotterdam, Netherlands: Sense Publishers.

Duschl, R., & Hamilton, R. (2011). Learning science. In R. Mayer & P. Alexander (Eds.), *Handbook of research on learning and instruction* (pp. 78–107). New York: Routledge, Taylor & Francis Group.

Eisenberg, N. (2010). Self-regulation and school readiness. *Early Education and Development, 21*(5), 681–698.

Flavell, J. H. (1979). Metacognition and cognitive monitoring: A new area of cognitive-developmental inquiry. *American Psychologist, 34*(10), 906–911.

Flavell, J. H., Miller, P. H., & Miller, S. A. (2002). *Cognitive development* (4th ed.). Englewood Cliffs, NJ: Prentice-Hall.

Furtak, E. M., Seidel, T., Iverson, H., & Briggs, D. C. (2012). Experimental and quasi-experimental studies of inquiry-based science teaching: A meta-analysis. *Review of Educational Research, 82*(3), 300–329.

Gallagher, S. (2000). Philosophical conceptions of the self: Implications for cognitive science. *Trends in Cognitive Sciences, 4*(1), 14–21.

Gibson, H. L., & Chase, C. (2002). Longitudinal impact of an inquiry-based science program on middle school students' attitudes toward science. *Science Education, 86*(5), 693–705.

Gunstone, R. F. (1994). The importance of specific science content in the enhancement of metacognition. In P. Fensham, R. Gunstone, & R. White (Eds.), *The content of science: A constructivist approach to its learning and teaching* (pp. 131–146). London, UK: Falmer Press.

Harris, K. R., Santangelo, T., & Graham, S. (2010). Metacognition and strategies instruction in writing. *Metacognition, strategy use, and instruction* (pp. 226–256).

Hodson, D. (2003). Time for action: Science education for an alternative future. *International Journal of Science Education, 25*, 645–670.

Keeley, P., & Tugel, J. (2009). *Uncovering student ideas in science: 25 new formative assessment probes* (Vol. 4). Arlington, VA: NSTA Press.

Keys, C. W., & Bryan, L. A. (2001). Co-constructing inquiry-based science with teachers: Essential research for lasting reform. *Journal of Research in Science Teaching: The Official Journal of the National Association for Research in Science Teaching, 38*(6), 631–645.

Krajcik, J., McNeill, K. L., & Reiser, B. J. (2008). Learning-goals-driven design model: Developing curriculum materials that align with national standards and incorporate project-based pedagogy. *Science Education, 92*(1), 1–32.

Krajcik, J., Reiser, B. J., Sutherland, L. M., & Fortus, D. (2004). *IQWST: Investigating and questioning our world through science and technology*. Ann Arbor, MI: University of Michigan.

Kramarski, B., & Mevarech, Z. R. (2003). Enhancing mathematical reasoning in the classroom: The effects of cooperative learning and metacognitive training. *American Educational Research Journal, 40*(1), 281–310.

Kuhn, D., & Dean, D., Jr. (2004). Metacognition: A bridge between cognitive psychology and educational practice. *Theory Into Practice, 43*(4), 268–273.

Lawson, A. E. (2010). Basic inferences of scientific reasoning, argumentation, and discovery. *Science Education, 94*(2), 336–364.

Martinez, M. E. (2006). What is Metacognition? *Phi Delta Kappan 87*(9), 696–699.

Metcalfe, J., & Kornell, N. (2005). A region of proximal learning model of study time allocation. *Journal of Memory and Language, 52*(4), 463–477.

National Academies of Sciences, Engineering, and Medicine. (2018). *How people learn II: Learners, contexts, and cultures*. Washington, DC: National Academies Press.

National Research Council. (2007). *Taking science to school: Learning and teaching science in grades K-8*. Washington, DC: National Academies Press.

National Research Council. (2012). *A framework for K-12 science education: Practices, crosscutting concepts, and core ideas*. Washington, DC: National Academies Press.

Roseman, J. E., Linn, M. C., & Koppal, M. (2008). Characterizing curriculum coherence. In 2061 connections (pp. 13–36).

Schneider, W., & Lockl, K. (2002). 10 The development of metacognitive knowledge in children and adolescents. *Applied Metacognition, 224*.

Schraw, G., Crippen, K. J., & Hartley, K. (2006). Promoting self-regulation in science education: Metacognition as part of a broader perspective on learning. *Research in Science Education, 36*(1–2), 111–139.

Shwartz, Y., Weizman, A., Fortus, D., Krajcik, J., & Reiser, B. (2008). The IQWST experience: Using coherence as a design principle for a middle school science curriculum. *The Elementary School Journal*, 151–171.

Sperling, R., Howard, B., Miller, L., & Murphy, C. (2002). Measures of children's knowledge and regulation of cognition. *Contemporary Educational Psychology*, 51–79.

The Gold Report. (n.d.). Palm Beach County Schools. Retrieved October 21, 2014, from http://www.palmbeachschools.org/dre/NRE/GoldReport.asp

Thomas, P. (2012). Metacognition in science education: Past, present and future considerations. In B. Fraser, K. Tobin, & C. J. McRobbie (Eds.), *Second international handbook of science education* (pp. 131–144). Dordrecht, The Netherlands: Springer.

Thwaite, A., & McKay, G. (2013). Five-year-olds doing science and technology: How teachers shape the conversation. *Australian Journal of Language and Literacy, The, 36*(1), 28.

Tusuzoglu, B. B., & Greene, J. (2014). An investigation of the role of contingent metacognitive behavior in self-regulated learning. *Metacognition Learning, 10*, 77–98.

Watts, M., Alsop, S., Gould, G., & Walsh, A. (1997). Prompting teachers' constructive reflection: Pupils' questions as critical incidents. *International Journal of Science Education, 19*(9), 1025–1037.

White, B., & Frederiksen, J. (1998). Inquiry, modeling, and metacognition: Making science accessible to all students. *Cognition and Instruction, 16*(1), 3–118.

White, B., & Frederiksen, J. (2005). A theoretical framework and approach for fostering metacognitive development. *Educational Psychologist, 40*(4), 211–223.

Whitebread, D., Coltman, P., Pasternak, D. P., Sangster, C., Grau, V., Bingham, S., … Demetriou, D. (2009). The development of two observational tools for assessing metacognition and self-regulated learning in young children. *Metacognition and Learning, 4*(1), 63–85.

Part III
Practitioner Inquiry: Lessons Learned From the Field of Science Teaching

Chapter 8
Lessons Learned and the Implications for Teacher Learning in Professional Development for Science Teachers

> The inquiry project had a major influence on my professional growth. The process allowed me to purposely reflect on my teaching and be intentional in addressing issues with my teaching and students learning. This has led me to be always learning and modifying my practices. The process is invaluable in helping me to zero in on issues of practice. I continue to use the inquiry process. I am looking at writing strategies now, as well as continuing reading strategies and have now added peer tutoring that emerged in response to questions in my first teacher inquiry project.

8.1 Professional Development and Teacher Learning

Professional development (PD) is about teachers' learning and improving their teaching practices to facilitate an increase in student achievement. The teachers in U-FUTuRES PD program, were being prepared to respond to the current reform efforts in science education. Practitioner inquiry was introduced to them as a tool to purposefully examine and make shifts in their teaching practices to accommodate the learning of the range of learners in their classrooms. In this chapter, I present an analysis of the teachers' responses to teacher inquiry one year after competing their initial practitioner inquiry. Data was garnered from interviews and teachers' formal reflection on their learning and the ways their practices were impacted by the knowledge constructed during the practitioner inquiry. From the analysis, I posit that practitioner inquiry should be added as one of the elements of effective PD. In this chapter, the historical background of reflection is discussed and positioned within the context of PD.

For each era of reforms in education, the nature of PD has been questioned, its activities analyzed for relevance and effectiveness, and research has sought to study and derive models of best practices. There is therefore no shortage of literature on PD in the general area of curriculum and instruction. In a longitudinal study conducted in 2002 by Desimone, Porter, Garet, Yoon, & Birman, they sought to answer the following question: What are the characteristics of professional development that affect teaching practice? In addition to a focus on content, knowledge, their findings confirmed five key effective features of PD identified in the national study National Academies of Sciences, Engineering, and Medicine (NASEM, 2015). Furthermore, they indicated that PD is more effective in changing classroom practices when there is collective participation and teachers engage in active learning

R. M. Pringle, *Researching Practitioner Inquiry as Professional Development*,
https://doi.org/10.1007/978-3-030-59550-0_8

opportunities, such as reviewing student work, and or obtaining feedback on their teaching. Other scholars while noting the consensus within the field regarding effective PD, caution that the list of important features is relatively extensive including a focus on clearly articulated and specific goals, with related activities and practices (Borko, Jacobs, & Koellner, 2010). However, as the research base continues to grow, educators agree that the field is not static, hence it is reasonable to expect more specificity over time (Borko et al., 2010; Stein, Smith, & Silver, 1999).

In addition, to the core features discussed in general curriculum and instruction, science education has included discipline specific components to the design and enactment of its PD.

These components include a focus on developing teachers' content knowledge in ways that will allow them deeper conceptual understanding, science-specific teaching practices, job-embeddedness and responsiveness to the local needs of science teachers (Loucks-Horsley, Stiles, Mundry, Love, & Hewson, 2010; Penuel, Fishman, Yamaguchi, & Gallagher, 2007; Wilson, 2013). The addition of the multiple components to science PD has impacted teacher beliefs and have augmented the instructional gains of the participants (Lotter, Smiley, Thompson, & Dickenson, 2016).

Furthermore, authentic science learning experiences that engage teachers in the skills and knowledge necessary for effectiveness have been shown to positively impact teacher beliefs and practices (Capps & Crawford, 2013; Lotter et al., 2016). Teachers' ability to teach science consistent with current reforms in science education is pertinent to workforce development and contributes to the nation's positioning in the global economic and knowledge marketplace. Many teachers engaged in PD may be new to reform-based practices and lack the knowledge and beliefs needed to achieve the vision of science education as presented in A Framework for K-12 Science Education and the Next Generation Science Standards (NGSS Lead States, 2013; NRC, 2012). Hence, PD in service to society becomes the process by which inservice teachers are afforded opportunities to build instructional capacity and transform science teaching and learning consistent with the vision of reforms.

The development of teachers as lifelong learners has a history as far back as Dewey in 1938. Dewey, at that time, observed and cited the burgeoning body of technology and scientific knowledge as impetus for teachers to keep pace by becoming continuous learners. Others have given credence to reflection as a tool that enables teachers to make careful considerations about their own experiences as they observe, frame and study problems within their own practices (Kayapinar, 2016; Zeichner & Liu, 2010). In this sense, teachers become reflective practitioners as they engage in a higher-order operation (Carr & Kemmis, 2009), which according to Dewey (1938), is a component of the action of solving the local problems related to teaching and learning. Schön (1987) used the term reflection-in-action to refer to the process by which thought-based actions are initiated and teaching and learning are viewed as continuous processes grounded in teachers' experiences. Schön sees the construction of tacit knowledge, knowing-in- action, occurring as experienced teachers in their everyday practices respond to and make the necessary adjustments

(Simon & Campbell, 2012). It is through this process of reflection that individuals recreate their actions and through critical practices enact relevant and appropriate actions (Grundy, 1987). Furthermore, this emerging knowledge, though not embraced in all spheres, does include theory, research, values and beliefs used by teachers to critically analyze and continually support the improvement of their teaching.

8.2 Researching Practitioner Inquiry

For the capstone project of the Master's degree in science education, teachers completed their practitioner inquiry in lieu of a thesis. They were fully engaged in conducting the practitioner inquiry beginning with identifying issues of their practices that were problematic in supporting students' learning and from which, guiding questions emerged. The teachers were introduced and scaffolded through the process of practitioner inquiry as they pursued the pathway of research – collecting and analyzing relevant data to respond to their questions. The practitioner inquiry as a non-linear process, occurred as the teachers were engaged in a cycle that included identification of questions, plan, observation and reflection and the enactment of improved practices. As the process developed, other relevant questions emerged and the cycle of inquiry continued. The project, as a degree requirement, culminated in the submission of a formal practitioner inquiry report that included a description of their teaching contexts, the nature of the inquiry, and discussion of the findings. The report also included a discussion of the implications for their practice going forward and a reflection on the process that addressed the development of skills to conduct the inquiry and their emerging posture toward the "stance of inquiry…a worldview and a habit of mind – a way of knowing and being in the world of educational practice," (Cochran-Smith & Lytle, 2009, p.vii).

The completion of the graduate degree was not the culmination of the PD program. At the end of the first semester, after the teachers were awarded the degree, they responded to a formal interview (see Appendix B, interview protocol), in which they explained the impact of the practitioner inquiry on their approach to teaching science, continuation of practitioner inquiry, attention to curriculum based on learning from their first inquiry and the extent to which practitioner inquiry could be considered an essential component of formal PD. Teachers' responses to the practitioner inquiry were also gleaned from the section in their capstone project in which they reflected on the process and their learning.

In writing their self-reflections, teachers were encouraged to be specific about the impact of the practitioner inquiry on their learning and on their pedagogical practices. Their reflections were guided by the following questions: What were some of the issues you faced as you conducted your practitioner inquiry? What have you learned about this research process and its impact on your teaching practices? In what ways did the practitioner inquiry impact your learning about yourself? In addition, the teachers were encouraged to provide justifications for any claims that were made.

Together, the interviews and the reflection afforded a qualitative data set and following Creswell and Poth (2017) data analysis process, both documents were read and coded with line-by-line descriptions followed by the identification of categories and themes.

8.3 Learning from Practitioner Inquiry

8.3.1 Impact on Teaching Practices

Teachers recognized the powerful influence of the practitioner inquiry on their teaching practices, their roles as classroom teachers, and their beliefs regarding their teaching and students' learning. Teachers described the process as profound; as impactful; and others characterized the process as transformative. Teachers expressed a broad variety of responses regarding the ways they were continuing to integrate and make changes to practices learned from their initial inquiry project. Teachers explained repeatedly how the practices learned during their practitioner inquiry changed their approach to teaching science, whether they continued to teach the same subject area or student population.

> I have never experienced a more significant or more powerful professional development tool than my teacher Inquiry. It transformed my classroom and the way my students achieve their understanding of science. It's not perfect, but my students are thinking for themselves and sharing with other students. I have students who are more enthusiastic and a classroom that is easier to manage. And, I have higher student achievement.

Practitioner inquiry provided a structured process for the teachers to examine and address the issues in their classrooms related to their practices and student learning. The process allowed them to feel empowered as they made changes to their teaching on behalf of their students learning. By systematically observing their classroom interactions, teachers recognized situations that they had previously thought were not under their control. For example, one teacher said that she was no longer willing to assume that there was nothing to be done for a struggling student. In the interview, another declared:

> I no longer write off a student or groups of students because factors in a student's life are not within my power or control. I might not understand, but I know how to research my teaching and I know that there are places I can go and it is ok to call my colleagues in the program about how I can address the issue in my teaching - they are people that I can ask as I now know that there are strategies I can try.

Teachers emphasized that they could try new approaches and assess improvement in students' outcomes without feeling as if they had to complete certain tasks within a given time and bracing for the bell. The practitioner inquiry and seeking to answer their questions forced them to be attentive to their practice and teacher learning rather than hurrying to satisfy curricular pacing guides. Teachers expressed a renewed awareness of the complexity of teaching and attending to the challenges of teaching and wanting to be effective. They were now open to seeing each lesson as a dynamic whole which was constantly shifting and required acute responsiveness and willingness to try. With teachers' awareness of the "unique and uncertain" patterns in their classrooms, they also realized that, through systematic efforts, they could make sense of aspects of their classroom practices and make changes. Teachers said they had learned to "monitor" students' using many formative assessments but the focus was now on their teaching and the many moves that occurred in any one teaching period. The act of being in the moment in their teaching allowed for the development of "a structure and format to address problem areas in my teaching." Another teacher noted that the inquiry project "made me look closely at student learning and the use of the formative assessments in making on the spot decisions."

8.3.2 The Unwelcome Truths of Teaching

For many of the teachers their Teacher Inquiries led them to what one researcher has described as "unwelcome truths" of the classroom (Mockler & Groundwater-Smith, 2015). As teachers addressed issues regarding equity, as they endeavored to serve struggling students from low-income areas and engage students from underrepresented populations they admitted some "unwelcome truths" about teaching. While they expressed discomfort, they recognized the importance of unearthing aspects of their teaching that were always not given the kind of focus needed to enhance and support the learning of all students. The following excerpt from a reflection on the process captures the group's feeling about being reflective and being more intentional about their teaching practices:

> Asking teachers to look at their own weaknesses is a difficult and sensitive process. Teachers have a hard time admitting their weaknesses. They are generally good people who want their students to succeed. But there are gaps between students, gaps because of gender or ability or race, and we need to be able to admit those gaps and adjust our practices. But, it has to be done in a non-punitive or evaluative way. It has to be done in a reflective way, and intentional so teachers can take a step back, look at their practices, analyse and understand the weaknesses that contribute to learning. When this is understood, then we could move forward with positive actions.

Teachers noted that such reflective practice was necessary to ensure equitable and effective instruction for all their students.

> In terms of enacting a reform-based science curriculum, it will not happen by simply putting the curriculum in the hands of the teacher and expecting results. It is simply not going to happen. The teacher has to engage in a self-reflective process to ensure that the curriculum is enacted in an equitable and effective way.

Some teachers struggled to admit that certain aspects of their classroom practices interfered with student engagement and learning. The practitioner inquiry brought to the fore certain implicit biases in their interactions with students from underrepresented populations of learners. It was an eye-opening experience for some teachers who were recognizing the possible biases in their teaching practices that disadvantaged students from underrepresented populations – seen through the lenses of low achieving students who often came from impoverished homes and had weak literacy skills. As one teacher expressed, "now I know…the ball is in my court to make a difference."

> It was very difficult as a person who considers herself to be all-inclusive, to admit that there was even a small bias in my practices. It is difficult to be able to take a step back and realize what these instances are and how to correct them. This is not something that all teachers will be comfortable with doing, even though it is a necessary part of the teaching to give all students a fair opportunity.

> Asking teachers to recognize bias is an incredibly difficult and sensitive process. Teachers do not want to admit that they may hold some bias, as they are generally good people who want all students to succeed. Admitting a bias is akin to admitting a weakness, or a character flaw. However, if we are going to succeed in eliminating the gaps between not only the genders, but the races and abilities, we need to be able to sit down and adjust our practice. I am now more cognizant of the participation of students, and find myself being more mindful that all students are participating and adjusting my teaching accordingly if this is not happening.

During their inquiries, they "took a step back" to look at their practices for students who were struggling science learners. For example, teachers were surprised to learn that they could integrate literacy skills into their science lessons.

> The students in my school are just more challenging. I have found that many of them have a difficult time in a lot of different areas. I wanted to do something with writing, coming up with ways to help these kids write. We are talking about kids who have totally different mindsets about science and math and the way that they approach learning is different. Their home life is different. What they bring to the table is different. It made all the difference in the world to me to have my own information about these kids that I collected and then use that information to work with them. And now I use my own information to work with my students. Every one of my students is dealing with their own different situation, but I am finding my way in teaching each of the students.

> I live in an area with a lot of student turnover. So, the students do not get much consistency which often translates into poor student performance. My inquiry was on writing instruction in science. What I did in my inquiry really gave me a tangible sense of the value of literacy instruction in content areas. It has made me more conscious of the literacy needs of my kids and it's made me better able to integrate literacy skills within the content. Once I had the knowledge of how to teach literacy skills, I could pull literacy into the inquiry piece. And that makes it more meaningful and accessible to the students.

Teachers explored a variety of strategies to engage their low achieving students and learned ways to meaningfully integrate strategies to address their learning needs. They recognized that the systematic and sustained focus became the key to understanding and reforming their teaching practices. Despite an increased awareness of their shortcomings, teachers also recognized the complexity of their work as teachers and the deep satisfaction when lessons went well such as seeing the changes in students who before were disengaged learners.

8.3.3 *Becoming Reflective Practitioners*

Prior to enrolling in U-FUTuRES, participants had thought of themselves as good teachers. They however, expressed they had more to learn and entered the program with the expectation of becoming better science teachers. As they reflected on the practitioner inquiry and the impact on their learning, they expressed surprise about how quickly their beliefs and practices became influenced by the intentional examination of their practices and the analysis of students' learning. One teacher's reflection captures the essence of this **element of surprise:**

> As I did the inquiry, I learned how imperfect I am, you know? Not that I ever thought I was perfect but the change took me by surprise quite honestly. I really thought I had it together and I watched myself teach, and I listened to myself teach. It was humbling. It was humbling as you realize how much you need to learn.... But the bottom line is I improved and that's the key thing. That is the whole deal with the inquiry and introspection. It is, the goal is to improve yourself.

Teachers explained that inquiry permeated their experiences in U-FUTuRES – both as a process for facilitating student learning in their science classrooms and as a process for facilitating their own continued learning as teachers. Many are continuing the practitioner inquiry focusing on other issues and questions that had surfaced about their teaching and to which they were seeking answers. They described how after a year later, the constant "wondering" about their practices was now a feature of their classroom interactions. The practitioner inquiry heightened their awareness making them more sensitive to shifts in students' responses, and, in the words of one teacher, "I find myself wondering about all kinds of things. Before the inquiry project, if I tried to analyze any type of student data … I did it in a really superficial way. Now I look at all data, whether its students' work, shifts in their level of engagement, or unexpected events. I am now more tuned in to focused reflection on what is occurring between my teaching and the students."

Teachers reported that conducting their Teacher Inquiries had inspired them to continue to reflect upon and improve practices for their students. They now defined teaching as a process that must include systematic reflection and intentional responses. The intentionality however, requires them to be ready to answer the following questions: What can we address? How can we make it better?

> As a teacher, you always hear about being reflective, it seemed sort of like thinking back and having nostalgic feelings about what did or didn't work. But the Inquiry Project forced me to be reflective in a way that I didn't even know I could be. [Doing the Inquiry Project] forced me to break down, to analyse, to pick apart, and then put back together everything that I had done. And that is a whole different way of looking at yourself as an educator. … Sometimes there is a tantrum thrown in there, too.

> As a teacher, I need to reflect on how what I am doing affects the students. There is not any point in being a teacher, if you are not willing to look back and say, what can we address? How can we make it better? Make the immediate change and then continually work towards further improving what you do.

Teachers said that their practices now included attention to their wonderings, systematic collection of student data, and making sense of what they have learned. The Inquiry Project provided the essential steps towards seeing the possibilities of observing and systematically raising questions, searching for answers, and making sense of phenomena in their classrooms. This was the essence of reflective practice leading to what one teacher described as, **"My classroom comes to life,"** and another noted, "I am now engaged in an "internal dialogue" that begins with my wondering about my students and what needs to be done as I become aware of patterns that were previously overlooked." One teacher noted, that she now had greater awareness of classroom interactions and events.

> Now, I look at students, and I go, what if I tried this or what if I tried that? So, I will record things when I am wondering. … Now, I have a guide to help me sort through all the things that were happening, things that I wasn't aware of. Those things come to life now, and I am able to improve as a teacher.

Others expressed during their interviews:

> When I began to analyze my practices, I became aware of patterns not seen before. I had a heightened awareness of my classroom, things that I had overlooked by daily trivialities or complicated problems that seemed to absorb my time. With the inquiry, I focused on goals I had for myself and my students. By reflecting on my practices, I try to accomplish what I hope my students are capable of doing. I want my students to be able to reflect, draw conclusions, and discover scientific principles from the evidence they have collected. In my inquiry, I was doing the same.

> Taking field notes and writing in the reflective journals were very valuable to me. I could actually see an increase in student participation which made me know for sure something was working. My teacher inquiry led to better student engagement and achievement. And I saw some real differences in the attitudes of the boys I was working with. So, that was a big takeaway for me. I was learning to be a better teacher.

Teachers said they now have multiple questions and ideas and wonderings to investigate. They were now immersed in the cycle of inquiry. They were now open to their teaching and purposeful in their observations, raising the kinds of questions that led to a period of focused work. As they became more systematic about finding answers, teachers recognized that learning about their practices was a continuous process.

> One result of this inquiry project is the realization that inquiry into my teaching never ends. If you are truly trying to become better, you have to understand that with inquiry, the more you know the less you really know. Each question leads to two questions. Each observation opens another door of observations. … Inquiry and introspection. They're almost interchangeable.

> I am constantly reflecting on what is successful and what's not successful in my classroom and evaluating what techniques are effective with the students based on those results. I reflect on my lessons a lot and then I plan for the next time I teach it. And I do that while I'm actually still in the moment, not when I pull up the lesson for next year. So that reflection has really had the biggest impact on me. So, I am constantly working with the moment and considering what changes could be made – what would be more beneficial for my students.

Teachers spoke eloquently about reflection during their inquiry process – as a method to further their inquiry and as an intense process of professional soul searching. As stated previously, Cochran-Smith and Lytle (2009) described the stance of inquiry as "a worldview and a habit of mind – a way of knowing and being in the world of educational practice" (p. vii). Clearly, the participants in this study were on their way to developing the stance of inquiry.

8.3.4 Practitioner Inquiry as Professional Development

When asked if practitioner inquiry should be a component of professional development for all science teachers, all teachers agreed that practitioner inquiry should be required for all science teachers. A few noted that it should be required of all teachers, regardless of content area or grade level. Participants responded, in part, by explaining how their inquiries benefitted their learning. It was clear from their responses that the practitioner Inquiry had far greater impact on their teaching practices than traditional models of PD, characterized by one teacher as the "sit and get" model of inservice education. In comparing previous experiences some teachers noted:

> Teachers tend to see typical PD as a day of wasted time. Someone comes from the district, shows us videotapes of perfect classrooms, tries to convince us to throw out our old ways and switch to the "new" reform and leaves us a bunch of handouts. Teacher Inquiry allows teachers to focus on their own needs instead of what an outsider perceives as the most pressing needs. But what's even more important is that teachers become learners and become self-developing rather than relying on others to lead the change.

> All the professional development we do … getting information from without … people from the district, people from other places, all that is useful. But in Teacher Inquiry, when you have teachers look at themselves, that is the best place to start.

The following interview responses sum up the essence of practitioner inquiry as a component of PD;

Teacher one:

> One problem that I have always found with PD is that I am listening to someone else's research and I'm listening to someone else's ideas and I always struggle with how to implement the topic in my own classroom. It doesn't necessarily work with my kids. When it comes to Teacher Inquiry, the information that you are able to collect from your own students, the data that you get as you try something different with them, that is much more effective to you as a teacher. You have collected your data on your own students and finding

> out what works best for your kids. ... To me, it's all the difference in the world to have my own information that I've collected and then to use my own information to work with my other students either that same year or the next years.

Teacher two:

> The inquiry project had a major influence on my professional growth. The process allows teachers to reflect on their practices in an intentional way to solve a challenge and then to actively modify your practices to meet that challenge. It's invaluable in helping teachers to foster changes in student learning and comprehension. I continue to use the inquiry process. I am looking at writing strategies now, as well as continuing reading strategies and peer tutoring from my project.

The nature of practitioner inquiry supported best practices, reflection on practices, and a process of sense making as teachers shaped their classrooms according to their learning and in response to the students within their own classrooms. The teachers stated that the experience of adapting best practices while monitoring students' achievement would be "invaluable" for other teachers. Thus, given their positive experiences with practitioner inquiry, participants had specific guidelines about how to provide effective practitioner inquiry as PD. They expressed that the effectiveness of the project and the success they achieved were directly related to the guidance and facilitation provided by the instructors of the practitioner inquiry course. "Learning to do inquiry is difficult," one teacher wrote but continued by expressing the need for initial support from knowledgeable educators and peers. She also suggested that practitioner inquiry should be introduced as one component of PD activities. This was consistent across the group – a belief that practitioner inquiry should be an element of PD for all teachers regardless of discipline.

> We took it step by step, coming up with a question, coming up with the kinds of data we need to answer this question, then what kinds of actions are you going to take? We had our cadre members to talk to and we took it step by step by step. How are other teachers going to see that happen? ... I think the support and the structure [we had from UF faculty] were vital to staying on track, and taking it step by step, and having a little bit of accountability and feedback to keep up with everything. Now I am on my own.

Teachers were advocating for a holistic approach to teachers' professional learning. One teacher noted that the U-FUTuRES Project had "inquiry" bookends. A course at the beginning of the program focused on reform-based science teaching in which students learned science as inquiry and in accordance with how scientists conducted science. At the end of the program, they engaged in practitioner inquiry, and as researchers, examined their teaching in response to questions asked. The process of inquiry served as a model for student learning in the teachers' classrooms and as the model for their professional development as teachers. One teacher explained that she modeled aspects of inquiry for her students as she collected data in her classroom.

> The students would see me writing in my journal when we would do things. I would tell them that I was collecting data and doing research as well, not on the science project, but keeping data about my [lesson] plans and their engagement and what they were doing with the science and what was not working. It's important for kids to understand that science is about research, it is about studying, it is about keeping records, it is about asking questions about when something works or when it doesn't. It all fits together.

Teachers noted that practitioner inquiry will ultimately lead to habits of inquiry as a part of their continued professional practice. They also agreed that it was valuable in empowering them to actively seek to improve their practices based on their own evidence. Nevertheless, they expressed several limitations. Practitioner inquiry should be conducted over an extended time, with adequate and expert scaffolding and support. The time, however, must be well organized, and teachers must be committed for the long term.

Teachers rated the influence of the practitioner inquiry on their (a) science learning, (b) teaching practices, and (c) stance as a researcher. Out of a total of five, science learning achieved a rating 3.8, their responses however varied widely, depending on their prior education. Teachers who had stronger background knowledge in science, tended to give a lower rating to this item. Teachers consistently gave the highest rating to impact on their teaching practices. As noted previously, regardless of their entering beliefs about their competence as teachers, the inquiry compelled them to re-evaluate and reform their practices. Some teachers viewed themselves as more discerning consumers of educational research and more capable of addressing research issues regarding data collection and analysis in their classrooms and school systems. Others saw themselves as more competent in engaging in research into their own practices.

Overall, teachers believed that the biggest changes were in embracing practitioner inquiry as research, unearthing the nuances in their classrooms such as implicit biases and assuming a stance of inquiry toward their teaching. Furthermore, their definition of reflective practice became more defined and rigorous, more thoughtful, and systematic bringing new insights and rewards to their practices.

8.4 Discussion

The teachers in this PD were engaged in a systematic and intentional study of their practices. What emerged were a series of learning related to their science-specific practices, student learning, and the critical examination of their lived experiences in the culture of science teaching in their schools. An analysis of the issues pertinent to this group of teachers revealed that the practitioner inquiry provided them a structured, yet powerful process to make sense of their classrooms and to feel empowered to make changes on behalf of their students. By systematically observing students, teachers explored possibilities in situations that they had previously thought were not under their control. One teacher said that she was no longer willing to assume that there was nothing to be done for a struggling student. For many participants in the study, their teacher inquiries led them to what one researcher has described as "unwelcome truths" of the classroom (Mockler & Groundwater-Smith, 2015). The practitioner inquiry provided an ease to ensuring equitable and effective instruction for all their students. The problems of practice and the "wondering" of teachers about their students' learning led to the realization that there were many reasons for marginalization in science learning. One such reason emerged in the practitioner inquiry as teachers addressed discrepancies in their practices and sought to be more effective in offering equitable science learning opportunities.

Teacher learning occurs within a complex system and when framed as causal reveals the conditions required for conceptual change (Opfer & Pedder, 2011). Consistent with current understanding about how learning occurs, our approach to PD activities was grounded in the teachers' classroom practices and aligned with the constructivist framework. The focus was also on providing a learner-centered structure that is responsive to the immediate needs of the teachers and their contexts. Here science teachers learn to teach and learn science in ways that may be substantially different from how they were taught. Providing inservice teachers the opportunity to engage in continuous professional learning not only requires a structure of supportive learning activities but also a set of "tools to develop the skills and practices of systematic, purposeful inquiry and critical reflection," (Hammerness, Darling-Hammond, Grossman, Rust, & Shulman, 2005, p. 437).

In the new paradigm for PD, teacher learning is considered a complex system. Teacher learning becomes enhanced as a result of multiple opportunities to learn, practice, and interact against the background of their previous experience (Johnson, 2006; Loucks-Horsley et al., 2010) and their perceived need for changes. The development of U-FUTuRES was shaped by the core features and the research base that contributed to the consensus model of effective PD (Borko et al., 2010; Desimone, 2009; Loucks-Horsley et al., 2010; NASEM, 2015). The elements of effective PD as described in the literature and mentioned throughout this book were clearly evident in the enactment of the program. As a comprehensive program, it involved key components supported by a range of activities that introduced teachers to robust theoretical bases and allowed the construction of new knowledge and practices through reflective dialogue and interactions with their peers, who as learners were all participants in the community of learners.

During the practitioner inquiry, teachers recognized the prominent place of reflection in their learning. Even though some held on to the idea that reflecting was a normal practice of teachers, they quickly recognized the impact of the systematic, and focused nature of the reflective practice in mediating their learning about their practice. As teachers became further entrenched in the process, a critical change was the development of a more questioning approach and a willingness to take risks. These risks involve taking actions they had never done before, such as examining their practices and sharing their experiences in ways that would also reveal vulnerability. Though the process began as mere description or recording of their observation, over time, they became more analytic as they responded to questions raised during their observations. Questions such as; what does this mean? What can be done differently?

Effective professional development programs provide teachers with opportunities to practice and reflect on new instructional strategies, to facilitate student thinking and student work, and to analyze examples of the target instructional practices.

Recent research investigating the conditions under which science teachers can succeed, suggests that focusing only on professional development to improve content knowledge and teaching skills – often referred to as human capital – may not be enough. I propose that practitioner inquiry be considered another element of PD. An element that has the potential to facilitate teachers' learning while securing their role as practitioner inquirer. However, teachers will need to construct knowledge about the process and procedures. Like any new concept, learners' prior ideas will need to be surfaced and through authentic experiences, construct knowledge about the process. This was the essence of the aspect of this PD that introduced teachers to practitioner inquiry. As practitioner inquirers, teachers were prepared and are now driven by the need to continuously focus on improving their practices by confronting their daily realities and making the necessary pedagogical adjustments.

8.5 Conclusion

The book describes a number of powerful ideas related to PD that were translated into practice during a federally funded program. Teachers were taught about the process of practitioner inquiry as one component of the graduate program. There were, however, two goals to conducting the practitioner inquiry – to satisfy the university's requirement as a capstone project for the graduate degree and teachers' development of the stance of inquiry in which, as educational researchers, would respond critically to problems of their practices. During the practitioner inquiry, teachers examined and reacted to their teaching in a cycle of activities that supported their continuous learning. In planning for and conducting the teacher inquiries, teachers focused on vexing issues in their own contexts that were negatively impacting students' learning. The process resulted in the building of closer and stronger relationships with their students, development of a range of teaching strategies specifically related to issues of learning, and as learners, a newly found level of empowerment.

References

Borko, H., Jacobs, J., & Koellner, K. (2010). Contemporary approaches to teacher professional development. *International encyclopedia of education, 7*(2), 548–556.

Capps, D. K., & Crawford, B. A. (2013). Inquiry-based professional development: What does it take to support teachers in learning about inquiry and nature of science? *International Journal of Science Education, 35*, 1947–1978.

Carr, W., & Kemmis, S. (2009). Educational action research: A critical approach. In S. E. Noffke & B. Somekh (Eds.), *Sage handbook of educational action research* (6th ed., pp. 74–84). London, UK: SAGE Publications Ltd.

Cochran-Smith, M., & Lytle, S. L. (2009). *Inquiry as stance: Practitioner research for the next generation*. New York, NY: Teachers College Press.

Creswell, J. W., & Poth, C. N. (2017). *Qualitative inquiry and research design: Choosing among five approaches*. Thousand Oaks, CA: Sage publications.

Desimone, L. M. (2009). Improving impact studies of teachers' professional development: Toward better conceptualizations and measures. *Educational Researcher, 38*(3), 181–199.

Desimone, L. M., Porter, A. C., Garet, M. S., Yoon, K. S., & Birman, B. F. (2002). Effects of professional development on teachers' instruction: Results from a three-year longitudinal study. *Educational Evaluation and Policy Analysis, 24*(2), 81–112.

Dewey, J. (1938). *Experience and education*. New York, NY: Macmillan.

Grundy, S. (1987). *The curriculum: Product or praxis*. London, UK: Falmer Press.

Hammerness, K., Darling-Hammond, L., Grossman, P., Rust, F., & Shulman, L. (2005). The design of teacher education programs. In L. Darling-Hammond & J. Bransford (Eds.), *Preparing teachers for a changing world: What teachers should learn and be able to do* (pp. 390–441). San Francisco, CA: Josey-Bass.

Johnson, C. C. (2006). Effective professional development and change in practice: Barriers science teachers encounter and implications for reform. *School Science and Mathematics, 106*(3), 150–161.

Kayapinar, U. (2016). A study on reflection in in-service teacher development: Introducing reflective practitioner development model. *Educational Sciences: Theory and Practice, 16*(5), 1671–1691.

Lotter, C., Smiley, W., Thompson, S., & Dickenson, T. (2016). The impact of a professional development model on middle school science teachers' efficacy and implementation of inquiry. *International Journal of Science Education, 38*(18), 2712–2741.

Loucks-Horsley, S., Stiles, K. E., Mundry, S., Love, N., & Hewson, P. W. (2010). *Designing professional development for teachers of science and mathematics*. Thousand Oaks, CA: Corwin Press.

Mockler, N., & Groundwater-Smith, S. (2015). Seeking for the unwelcome truths: Beyond celebration in inquiry-based teacher professional learning. *Teachers and Teaching, 21*(5), 603–614.

National Academies of Sciences, Engineering, and Medicine. (2015). Science teachers learning: Enhancing opportunities, creating supportive contexts. Committee on strengthening science education through a teacher learning continuum. Board on science education and teacher advisory council. In *Division of Behavioral and social science and education*. Washington, DC: The National Academies Press.

National Research Council. (2012). *A framework for K-12 science education: Practices, Crosscutting concepts, and Core ideas*. Washington, DC: The National Academies Press.

NGSS Lead States. (2013). *Next generation science standards: For states, by states*. Washington, DC: National Academies Press.

Opfer, V. D., & Pedder, D. (2011). Conceptualizing teacher professional learning. *Review of Educational Research, 81*(3), 376–407.

Penuel, W. R., Fishman, B. J., Yamaguchi, R., & Gallagher, L. P. (2007). What makes professional development effective? Strategies that foster curriculum implementation. *American Educational Research Journal, 44*(4), 921–958.

Schön, D. A. (1987). Educating the reflective practitioner. San Francisco: Jossey-bass Simon, S., & Campbell, S. (2012). Teacher learning and professional development in science education. In *Second international handbook of science education* (pp. 307–321). Dordrecht, Netherland: Springer.

Simon, S., & Campbell, S. (2012). Teacher learning and professional development in science education. In *Second international handbook of science education* (pp. 307–321). Dordrecht: Springer.

Stein, M. K., Smith, M. S., & Silver, E. (1999). The development of professional developers: Learning to assist teachers in new settings in new ways. *Harvard Educational Review, 69*(3), 237–270.

Wilson, S. M. (2013). Professional development for science teachers. *Science, 340*(6130), 310–313.

Zeichner, K., & Liu, K. Y. (2010). A critical analysis of reflection as a goal for teacher education. In *Handbook of reflection and reflective inquiry* (pp. 67–84). Boston, MA: Springer.

Appendices

Appendix A

U-FUTuRES Cohort 2 Cadre Meeting – Terrace Room – University of Florida September 12, 2013

Elements of Effective Science Instruction and how Children Learn

Learning Goal Discuss the elements of effective science instruction with specific representation of surfacing prior ideas and sense-making in IQWST lessons.

Surfacing: Think About <u>To what extent is it important that the pedagogy used in our teaching is consistent with the ways we believe learning takes place? – No response needed</u>

- What are the elements of effective instruction? STLs to write their ideas on index cards – Think – Write – Share with a partner

Activity # 1: Video Case Analysis: Individually STLs to observe a video clip of an IQWST lesson and instructed to document instances of <u>**surfacing students' prior ideas**</u> and <u>**sense making**</u>: Poster paper with <u>surfacing students' prior ideas</u> and <u>sense making</u> for display

- In groups, share and talk briefly about their observation
- As a group, write one claim about the presence of these two elements in observed portions of the lesson – support with evidence from observations

R. M. Pringle, *Researching Practitioner Inquiry as Professional Development*,
https://doi.org/10.1007/978-3-030-59550-0

What do we now believe about how learning occurs?
Debriefing – elicit the importance of past experiences in learning

– Establish that in *surfacing prior ideas* many strategies can be used (KWL, open-ended teacher questions, demonstration, students' questioning) but regardless of the strategy, in order to confront or build on students' initial thinking, they must bring their ideas to the surface so that they themselves have an awareness of their ideas – teachers to note opportunities for including/integrating Common Core standards such as encouraging students to engage in the writing of their ideas etc.

Activity # 2: Personal Lesson Sharing In groups of threes, share individual lessons focusing on the sense-making process. Within group discussion to include ideas and or suggestions to strengthen the sense-making component of the shared lessons. In what ways are you able to connect the sense-making to the learning goals of the lesson?

Summary input from lesson sharing:
Debriefing – Elicit and establish strategies, approaches, and importance of incorporating sense making in all science lessons – STLs to provide summary input from their discussions.

Activity # 3: Classroom Vignette Analysis: Provide Vignette, *When Rocks Collide***?**

Distribute the*Elements of Effective Instruction*:

1. STLs to use the criteria and questions to analyze the effectiveness of the lesson described in the vignette. Questions to guide vignette analysis:
   - How effective was the lesson in providing opportunities for achieving the learning goal?
   - What were some strengths of the lesson?
   - What were some issues?
   - If you had observed this lesson guided by the Elements of Effective Instruction as an observation protocol, what advice would you give to this teacher?
2. Write 3 claims and the accompanying evidence to support the claims.
3. Identify the IQWST features in the lesson

Note – Each person in the group to keep a record of the observation and the analysis made.

Activity # 4: Teachers to read real lesson from IQWST teacher edition

- What were some features of IQWST that were in the teacher handbook but not present/obvious in the lesson that was implemented?
- What are some possible reasons for the differences?
- In what ways did the differences impact the effectiveness of the lesson?

Whole group sharing of ideas from discussion (Real lesson):

Activity # 5: Sense-Making: Look back at your initial note cards – What are some elements of effective instruction?

What are some possible ways to incorporate and ensure that the sense-making element is included in ALL lessons?

Have you ever conducted science labs or activities with your students and somehow, they missed what they should have learned? How can the sense-making component reduce this occurrence?

Teachers to discuss responses in pairs.

Appendix B

U-FUTuRES: Follow-up Interview Protocol Practitioner Inquiry Projects

Questions

1. In what ways, if any, did conducting the study of your own practice (teacher inquiry project) influence your current approach to teaching science in your classroom?
2. Are there aspects of the process of teacher inquiry that you have adapted in your practice? (Explicitly or implicitly continuing to do) Explain.
3. Would you recommend that all science professional development include teacher inquiry-projects? Why or why not?
4. Reflect on the specific curricular area you focused on during your teacher inquiry project, are your findings making a difference in your teaching? In what ways?
5. What did you learn about yourself as a teacher from doing the teacher inquiry project?
6. Reflect on all activities in the entire PD program, which ones best supported your learning about reform-based science practice?

In the section below, supply a number and then provide an explanation to support your response.

1. On a scale of 1–5 with 1 being no impact and 5 being greatest impact, how did engaging in the teacher inquiry project support your **science learning?** Explain
2. On a scale of 1–5 with 1 being no impact and 5 being greatest impact, how did engaging in the teacher inquiry project impact your **teaching practices?** Explain
3. On a scale of 1–5 with 1 being no impact and 5 being greatest impact, how did engaging in the teacher inquiry project support your **stance as a researcher?** Explain

The manufacturer's authorised representative in the EU is Springer Nature Customer Service Centre GmbH, Europaplatz 3, 69115 Heidelberg, Germany. If you have any concerns regarding our products, please contact ProductSafety@springernature.com

Printed and bound by CPI Group (UK) Ltd, Croydon, CR0 4YY
30/09/2026
02225453-0010